The Electronic Commerce Dictionary

The Electronic Commerce Dictionary

The definitive terms for doing business on the Information Superhighway

by Ted Haynes

THE ROBLEDA COMPANY

MENLO PARK, CALIFORNIA

The Electronic Commerce Dictionary is published by:

The Robleda Company
1259 El Camino Real, Suite 2720
Menlo Park, CA 94025 U.S.A.

Design and Production by Joel Friedlander Publishing Services

Available through book wholesalers. ISBN 0-9646506-0-6

First Printing
Printed in the United States of America
10 9 8 7 6 5 4 3 2 1

Publishers Cataloging in Publication
(Prepared by Quality Books Inc.)
Haynes, Ted.
    The electronic commerce dictionary : the definitive terms for doing business on the information superhighway / Ted Haynes.
    p. cm.
    Preassigned LCCN: 95-69169
    ISBN 0-9646506-0-6

    1. Business enterprises—Communication systems—Dictionaries. 2. Internet (Computer network)—Dictionaries. 3. Electronic data interchange—Dictionaries. 4. Electronic funds transfer—Dictionaries. 5. Computer security—Dictionaries. 6. Home banking services—Dictionaries. I. Title.

HD30.335.H39 1995   650'.028'5'467'03
                   QBI95-20229

▪ Contents ▪

v

▪ Acknowledgements ▪

This book would never have seen the light of day without the assistance of several important people. First of all the author would like to thank Len Feldman, author of *Windows NT, The Next Generation* for editing the manuscript and for his excellent advice throughout the project. Many thanks to those who reviewed the manuscript and provided comments more valuable than they perhaps knew: Dave Darnell, Dieter Karaluz, Paul Hoffman, Glenn Fleishman, and Michael Killen. Any remaining errors are strictly the responsibility of the author. For encouraging the author to explore the Internet and to write this book thanks to Jared Haynes of the Department of English at UC Davis, Isabel Haynes, Ann Zeichner, Tom Parker, and Bob Metcalfe. Heartfelt thanks to my wife, Joan Haynes, for her support and patience with my long hours at the PC and on the Internet.

Finally, a thanks in advance to you, the reader, for any comments, criticisms, or new definitions you provide. The author can be reached at T_Haynes@ix.netcom.com

▪ Introduction ▪

Whatever you cannot understand you cannot possess.
— *Goethe*

Every revolution in technology and business brings with it a host of new words, abbreviations, and acronyms. Electronic commerce is spawning more than its share—digital cash, electronic catalog, Internet presence, network payment system—to name a few. Managers are finding they need to learn the nomenclatures of the Internet, electronic data interchange (EDI), electronic funds transfer (EFT), bank card operations, and many other fields covered in this book. Driven most recently by the growth of the World Wide Web, electronic commerce is poised to change the business landscape permanently.

This book is for anyone working to understand electronic commerce and looking for opportunities to profit from it. The definitions are based on usage in both general and technical sources—books, seminar materials, government publications, magazines, over two hundred World Wide Web sites and FTP servers, and the author's years of experience in commerce and data communications. The intent has been to make each term understandable to anyone with modest computer literacy. Where appropriate, specific products and companies are identified.

Cryptography, long the purview of government agencies and the military, turns out to be the bedrock on which many new ways of doing business over public networks are founded. Although many cryptographic terms are defined in this book, the relatively sophisticated math behind RSA and other algorithms is not described in detail. There are several excellent books on network security and on cryptography which can provide in-depth information.

Where words or phrases important to a definition are themselves defined elsewhere they are printed in SMALL CAPITAL LETTERS. Plural words in capitals (e.g. MICROPAYMENTS) are usually defined in the singular form. Where two or more words are in capitals (e.g. WEB BROWSER) the words will sometimes be defined separately rather as a single phrase.

The dictionary is designed to remain a fundamental resource in the development of electronic commerce. New words and other updates to this book will be published on the Electronic Commerce Dictionary Web site at http://www.haynes.com/haynes. Readers are encouraged to check this site for new information periodically.

The Electronic Commerce Dictionary

· A ·

acceptable use policy the policy of an INTERNET SERVICE PROVIDER on what uses of the network are permitted. For example, many ISPs specifically forbid chain letters. The NATIONAL SCIENCE FOUNDATION, which once provided Internet BACKBONE services, limited backbone services to research and educational institutions and to open scholarly communication and research by companies.

acceptor The party, such as a merchant, who accepts a PAYMENT CARD in exchange for goods or services and presents the TRANSACTION data to the ACQUIRER.

access Entry to a computer system or network, such as the INTERNET.

access control Protection against unauthorized use or manipulation of RESOURCES.

ACH See AUTOMATED CLEARING HOUSE.

ack Notification sent from one device to another to acknowledge that a MESSAGE has been received.

acquirer The institution, such as a BANK, which receives PAYMENT CARD TRANSACTION data from an ACCEPTOR and passes it on to the authorizing institution (the card ISSUER). Also known as the MERCHANT BANK or merchant processing bank.

Acrobat A software product from Adobe Systems that modifies FILES for transfer between incompatible computers such that the file will be displayed and printed identically on both machines.

Advance shipping notice (ASN) Also known as the ship notice/manifest, an EDI STANDARD TRANSACTION SET sent by a vendor to a customer specifying the contents of a pending shipment and the estimated time of arrival.

Advanced Intelligent Network A switched network using sophisticated software in which the information carried is separated from the signaling and control signals. The current public switched telephone network implements some elements of advanced intelligent networking.

Advanced Research Projects Agency (ARPA) An organization within DOD which developed the predecessor to the Internet (ARPAnet) and which is now chartered with identifying and supporting risky technologies with significant long-term benefits. Prior to 1993, it was known as DARPA for Defense Advanced Research Projects Agency.

Advanced Technologies Program (ATP) A program administered by NIST which grants funds to small companies to undertake research on generic technologies prior to their becoming viable in the marketplace.

agent A software program that processes queries and sends responses on behalf of an APPLICATION.

AIAG See AUTOMOBILE INDUSTRY ACTION GROUP.

American National Standards Institute (ANSI) The U.S. standardization body that administers a wide variety of standards, including the X12 standard for EDI. ANSI is a member of the INTERNATIONAL STANDARDS ORGANIZATION.

American Textile Partnership (AMTEX) A CRADA formed by the textile industry and the Department of Energy.

AMTEX See AMERICAN TEXTILE PARTNERSHIP.

anchor In a WORLD WIDE WEB page, a word or words that are highlighted on a screen or displayed in a different color from other text to indicate a LINK to another point on the same page or on another page. A USER can elect to display the other point or page by clicking on the anchor. See HYPERTEXT.

ANI See AUTOMATIC NUMBER IDENTIFICATION.

anonymity The ability to send a MESSAGE or pay out funds without revealing the IDENTITY of the sender or payer.

anonymous FTP An Internet tool that enables user ACCESS to a REMOTE HOST's directories and FILES without requiring a pre-arranged PASSWORD See FTP.

anonymous server A SERVER which enables a USER to send an anonymous POSTING to USENET by deleting all user-related information and forwarding the MESSAGE to the USENET newsgroup.

ANSI See AMERICAN NATIONAL STANDARDS INSTITUTE.

ANSI ASC X12 The ANSI accredited subcommittee responsible for developing the EDI document standards. See X12.

API See APPLICATION PROGRAM INTERFACE.

application A software program which performs tasks directly useful to an individual or organization, as opposed to an operating system, software tool, driver, or utility which supports other programs or is useful only in manipulating computer systems, devices, or networks.

application acknowledgment An EDI STANDARD TRANSACTION SET that acknowledges receipt of an earlier transaction set and its processing by a business APPLICATION.

application level gateway A FIRE WALL technology that uses a single device with only one IP ADDRESS to represent a private network to users on the Internet. It provides stronger SECURITY than PACKET LEVEL FILTERING but is more difficult to implement. An application level gateway can be a PROXY SERVER and can support IP address translation, user AUTHENTICATION, and end-to-end ENCRYPTION.

application program interface (API) A set of rules stating how an APPLICATION program calls a utility or other software program. In practice, a set of semi-standard interfaces between two applications or between applications and an operating system.

appropriate usage policy Rules set by a company on how employees should use company networks and the Internet, particularly with regard to E-MAIL, and concerned with such issues as use for private gain, soliciting donations, representation of company positions, offensive, harassing or disparaging statements, sexually oriented images or messages, transmission of proprietary information, and COPYRIGHTS.

Archie An INTERNET application for locating FILES by name. See VERONICA and JUGHEAD.

ARPA See ADVANCED RESEARCH PROJECTS AGENCY.

arrival notice An EDI STANDARD TRANSACTION SET sent from an ocean carrier to an onshore carrier and to the consignee to report on the planned arrival of shipped goods.

ASCII American Standard Code for Information Interchange. An 8-BIT code (including one bit for parity) for representing 256 characters of the English alphabet, foreign language characters, numbers, punctuation, and selected symbols.

ASN See ADVANCE SHIPPING NOTICE.

asynchronous transmission The TRANSMISSION of BITS over a NETWORK without precise clocking and with no common time reference between the sender and receiver. Individual characters are normally encapsulated in control bits called start and stop bits. See SYNCHRONOUS TRANSMISSION.

Asynchronous Transfer Mode (ATM) A technology which permits the sharing of TRANSMISSION facilities among data types with different requirements (e.g. voice, data, video) by carrying the data in small fixed-size (56 byte) cells. The ITU-TSS standard for CELL RELAY.

ATM See ASYNCHRONOUS TRANSFER MODE, AUTOMATED TELLER MACHINE.

ATM card See AUTOMATED TELLER MACHINE.

ATM network A SYSTEM that allows customers to use AUTOMATED TELLER MACHINES owned and operated by BANKS and financial institutions other than their own. ATM NETWORKS are jointly operated by multiple owners and operators of ATMs and by a network service provider, who provides ELECTRONIC FUNDS TRANSFER.

ATP See ADVANCED TECHNOLOGIES PROGRAM.

authentication 1) In accessing a computer system or network, methods of assuring that the party requesting ACCESS is in fact the party he or she claims to be. See SECURITY, PASSWORD. 2) In E-MAIL and ELECTRONIC COMMERCE, assurance that a party to a TRANSACTION or communication is the person or organization so represented and that any MESSAGE sent by that party has not been modified in transit. See DIGITAL SIGNATURE and DIGITAL CERTIFICATE. 3) In EDI, methods of assuring that any changes in a document during TRANSMISSION can be detected by the receiving party. 4) In EFT, a method of assuring that a payment instruction has indeed originated at the sending BANK and has not been tampered with.

authentication key In EDI, a character sequence that both parties to a TRANSACTION must use to begin the AUTHENTICATION procedure. See KEY.

authorization 1) The approval to complete a TRANSACTION. In BANK CARD transactions authorization is given by the ISSUER to the ACQUIRER. 2) The granting of rights to users to ACCESS resources or to read, modify, delete, copy, classify or otherwise manipulate specific information.

authorizing key A set of BITS which must be present for an encrypted software program to run.

automated clearing house (ACH) Any of several organizations which BANKS use to settle their accounts with one another electronically as an alternative to FED WIRE. Most Social Security payments and payroll direct deposit transactions are handled through the AUTOMATED CLEARING HOUSE network. ELECTRONIC BENEFITS TRANSFER is a controversial potential use of ACH networks. See also NATIONAL AUTOMATED CLEARING HOUSE ASSOCIATION.

automated teller machine (ATM) A machine which enables a BANK customer using a secret PERSONAL IDENTIFICATION NUMBER and a plastic DEBIT CARD with a magnetically encoded strip (an ATM

CARD) to transact business with the bank, including the withdrawal of cash, at any hour of the day or night. See also ATM NETWORK.

Automatic Number Identification (ANI) A telephone company capability, available nationwide, that passes the telephone number of a person calling an 800 number to the called party.

Automobile Industry Action Group (AIAG) The X12 standard for EDI as modified for use by the automobile industry. Also, the committee that maintains the modified standard.

autonomous system A group of networks administered by a single authority (e.g. a company wide area network).

availability The assurance that legitimate users of a RESOURCE or information are not denied ACCESS to it.

▪ B ▪

backbone A NETWORK that connects other, usually lower BANDWIDTH networks, and allows them to send data to one another.

bandwidth The amount of data that can be transmitted over a circuit or NETWORK, measured in BITS per second.

bank A company that maintains savings and checking accounts, issues loans and credit, and deals in securities issued by governments and corporations. A commercial BANK invests primarily in loans while an investment bank invests in securities for its CLIENTS and for its own account. See also FINANCIAL INSTITUTION.

bank card TRANSACTION CARD that enables a BANK customer to purchase goods and services and/or receive cash at AUTOMATED TELLER MACHINES. Bank cards include credit cards, and debit cards.

bank card association An organization, owned by financial institutions, that licenses and markets BANK CARDS and service marks, facilitates TRANSACTION AUTHORIZATION, and handles accounting and transferring of funds between BANKS in connection with bank card transactions. The two associations in the United States are MasterCard International and Visa (USA).

bar code a printed array of bars and spaces of varying width corresponding to an alphanumeric sequence that provides information about the item displaying the bar code. By means of the bar code the alphanumeric data can be quickly input into a computer using a SCANNER. Bar codes are widely used to track inventory and shipments. Bar codes on most consumer items conform to the UNIVERSAL PRODUCT CODE (UPC).

Basic Rate Interface (BRI) The ISDN interface designed for individual subscribers and consisting of two channels for voice or data and one channel for signaling.

BBS See BULLETIN BOARD SYSTEM.

bill of lading An EDI STANDARD TRANSACTION SET sent from a shipper to a carrier or consignee to detail the contents of shipment and delivery details of a shipment ready to be picked up.

BINHEX A Macintosh compressed FILE format used to transmit files over a network such as the Internet.

biometric authentication Verification of a person's IDENTITY by comparison of a unique physical characteristic (e.g. fingerprint, retinal scan) with a previously verified measurement.

BISNIS A Department of Commerce network that provides ONLINE information on business opportunities in the former Soviet Union.

bit The smallest unit of information. A binary digit equal to either zero or one.

BITNET A network between universities in the U.S., Europe, and Japan which does not use TCP/IP but which can exchange E-MAIL with the Internet.

bits per second (bps) A measure of the speed of any digital TRANSMISSION system.

blind entry In DIRECT STORE DELIVERY the acceptance of a shipment and updating of inventory information without an actual count of the units received.

BOL See BILL OF LADING.

book transfer The transfer of funds from one account to another within the same FINANCIAL INSTITUTION.

booking confirmation An EDI STANDARD TRANSACTION SET relaying acceptance of a freight booking on an ocean carrier.

booking request An EDI STANDARD TRANSACTION SET providing the details of a request to place freight with an ocean carrier.

bot A computer program whose output appears to be the work of an ONLINE human being, sometimes replying to general E-MAIL inquiries (a mailbot) and sometimes assuming a personality and taking on the role of a player in an online game. See KNOWBOT.

BRI See BASIC RATE INTERFACE.

bridge A device which links two local area networks and forwards packets between them, filtering out packets addressed to computers on the LAN from which the packets originated.

broadband Any TRANSMISSION system which combines multiple signals on a single physical circuit, as does cable TV. See ASYNCHRONOUS TRANSFER MODE.

browser A CLIENT program that facilitates locating and displaying information. The most popular WORLD WIDE WEB browser is Netscape, based upon MOSAIC.

bulletin board system (BBS) A computer system accessible over the Internet or by direct dialup or both that provides news and information on a particular topic and frequently supports POSTING of MESSAGES and news by its subscribers. Though BBS software is not standardized, functionality is similar across BBS software packages.

· C ·

CA See CERTIFICATION AUTHORITY.

cable area network A multi-star topology for cable TV that runs fiber-optic trunks to selected amplifiers in the network and conventional coax cable to end users. The fiber star provides improved signal quality and reliability versus the older tree-and-branch cable topology. About 25% of cable TV subscribers are served by these fiber-enriched systems.

cable modem A MODEM that provides ACCESS to on-line computer services and the Internet delivered over a cable TV system.

CAD See COMPUTER-AIDED DESIGN.

CAD/CAM computer-aided design/computer aided manufacturing.

CAFE Conditional Access for Europe. A group cooperating to develop a secure electronic PAYMENT SYSTEM using SMART CARDS and ELECTRONIC WALLETS that protects the PRIVACY of the USER. Members include Siemens, France Telecom and Post Research, and DIGICASH. See DIGITAL CASH.

caller ID services Services provided by telephone companies that enhance SECURITY by providing the phone number of the calling party over the phone line to the receiving party before the call is picked up. Variations of the services are Caller Number Delivery (CND), and Calling Line Identification. Services are available only where permitted by local regulation. The major drawback to the service is that the information is not passed between phone carriers and cannot be used in granting ACCESS to a computer when users call through a phone carrier other than that serving the computer itself.

Canadian Payments Association (CPA) An industry association created by Parliament that operates a national CLEARINGS and

SETTLEMENTS system and plans the evolution of the national PAYMENTS SYSTEM. The CPA operates the Automated Clearing Settlement System (ACSS) and the US Dollar Bulk Exchange (USBE) system.

Capstone A U.S. government project to develop publicly available CRYPTOGRAPHY standards incorporating KEY ESCROW and the DECRYPTION of encoded messages by government agencies under court order. The four major components of Capstone are SKIPJACK (a data ENCRYPTION algorithm), CLIPPER (a chip incorporating the Skipjack algorithm), DSS (a DIGITAL SIGNATURE algorithm), and SHS (a HASH FUNCTION). All parts of Capstone have 80-BIT SECURITY and all the KEYS involved are 80 bits long.

carbon copy routing the sending of duplicate EDI standard transaction sets to more than one computer.

card-not-present transaction CREDIT CARD or DEBIT CARD TRANSACTION, such as a catalog order over the phone or over a network, where the vendor cannot check that the buyer has physical possession of the card itself.

CARI A system of making CREDIT CARD purchases through the Internet without sending credit card numbers over the network. The customer sends the merchant a virtual credit card number assigned by CARI and the merchant sends the number to CARI along with the buyer's encrypted telephone number and a preferred time to call. When contacted, the customer enters into the phone keypad the virtual credit card number, a PIN, and the credit card number and expiration date of a credit card. CARI combines this with the order information from the merchant and formats a transaction file.

carpet bomb To CROSSPOST an inappropriate MESSAGE (e.g. an advertisement, a chain letter, a misleading business solicitation, or an intentionally offensive message) on USENET or Internet newsgroups.

CASE Computer Aided Software Engineering.

cash card An AUTOMATED TELLER MACHINE DEBIT CARD.

cash concentration and disbursement (CCD) An AUTOMATED CLEARING HOUSE standard transaction set for making payments electronically.

CBDS See CONNECTIONLESS BROADBAND DATA SERVICE.

CCD See CASH CONCENTRATION AND DISBURSEMENT.

CCD+ A CCD transaction set which permits the addition of free form text.

CCITT Consultative Committee for International Telegraph and Telephone. Now named the International Telecommunication Union - Telecommunication Standardization Section or ITU-TSS.

CD ROM Compact Disc Read Only Memory. A prerecorded optical data storage format using the same media as digital audio Compact Discs.

CDMA See CODE DIVISION MULTIPLE ACCESS.

CDPD See CELLULAR DIGITAL PACKET DATA.

CEBus An architecture for managing energy in the home and controlling appliances backed by the Electronic Industry Association using an open communication system based on OSI.

CellNet A system of using POWER LINE CARRIER to send signals from SMART METERS to neighborhood radios which each transmit to a single central site for billing by a utility. Also see NETCOMM.

cell relay A NETWORK technology simultaneously carrying all types of transmissions (voice, data, video, etc.) in cells of identical size. The fixed-length of the cells makes it possible to switch each cell to its appropriate circuit at very high speeds.

Cello A popular WWW BROWSER.

Cellular Digital Packet Data (CDPD) A technology which uses the same radio channels as analog cellular voice services to provide data networking services to mobile hosts.

CERN The European Laboratory for Particle Physics at which Tim Berners-Lee originated the WORLD WIDE WEB and which joined MIT in developing international standards for the Web.

CERT Computer Emergency Response Team. An organization established at Carnegie Mellon University to respond to any sudden wide-ranging problem on the Internet, particularly SECURITY related issues.

certificate issuing authority An organization which issues DIGITAL CERTIFICATES. To be useful, a certificate issuing authority must be widely known and trusted, and must have well defined methods of assuring the IDENTITY of the parties to whom it issues digital certificates.

certificate issuing system A hardware and software system which issues, manages, and reports on serialized DIGITAL CERTIFICATES.

Certificate Revocation List (CRL) A list of digital certificates which have been revoked or held in suspension.

Certification Authority Any of the Level 3 organizations in the PRIVACY ENHANCED MAIL model which is certified by a POLICY CERTIFICATION AUTHORITY and provides certification to users of the certificates.

certified delivery Provides a document to the vendor of DIGITAL MERCHANDISE that proves that the merchandise has been received by the purchaser over the network.

CGI See COMMON GATEWAY INTERFACE.

challenge/response authentication A method of authenticating remote users through an ENCRYPTION algorithm and a TOKEN embedded in software or in a hand-held device or via a series of questions that only the expected remote user can answer. Also see TWO-FACTOR AUTHENTICATION.

CheckFree An automated PAYMENT system which allows consumers to make payments from their checking or credit card account by sending AUTHORIZATION to CheckFree through a dialup port. CheckFree arranges the transfer of funds from the consumer's bank to the merchant by ELECTRONIC FUNDS TRANSFER. The advantage to the consumer is easier payment and record keeping. For the merchant, the use of EFT means the funds are available for use the day after collection (sooner than if they waited for check CLEARING through the banking system).

Chemical Industry Data Exchange (CIDX) Guidelines for use in the chemical industry of the XI2 standard. Also, a group that manages EDI activities for the chemical industry.

CIDX See CHEMICAL INDUSTRY DATA EXCHANGE.

CIE See CUSTOMER INITIATED ENTRY.

cipher A cryptographic algorithm - a mathematical function for ENCRYPTION and DECRYPTION.

cipher text An encrypted MESSAGE.

circuit-switched network A NETWORK, such as the telephone system, which establishes a physical circuit between two or more parties in order to transmit information between them.

CIX See COMMERCIAL INTERNET EXCHANGE ASSOCIATION.

clearings The delivery of checks from BANKS where they have been deposited to the banks on which the checks were written, and the movement of funds in the reverse direction. Some paper checks are cleared through the U.S. Federal Reserve check clearing system and some are cleared through clearing houses owned by banks. Also see ELECTRONIC FUNDS TRANSFER.

cleartext See PLAINTEXT.

client A software program or, loosely, a person or computer, that obtains services (e.g. E-MAIL, FTP, WORLD WIDE WEB) from servers on the Internet. More generally, any computer or APPLICATION that requests services or information from another computer or APPLICATION. See CLIENT/SERVER COMPUTING.

client/server computing Cooperative computing between two computers or APPLICATIONS using extensive computer-initiated communication between the CLIENT (which requests services) and the SERVER (which responds to the requests).

Clipper Within the U.S. government's CAPSTONE project, the CLIPPER chip implements the classified SKIPJACK algorithm. The term Clipper is often used to refer to the Skipjack algorithm.

CNRI Corporation for National Research Initiatives.

Code Division Multiple Access (CDMA) A cellular technology developed by Qualcomm that allows 100% frequency reuse from cell to cell because all the frequencies can be used in each cell and cells can be made smaller as needed.

com The top-level Internet DOMAIN NAME assigned to all for-profit organizations. In the United States, a business domain name almost always ends with a period followed by com (eg. 3Com.com).

COMM ID In the grocery industry, an identifier placed within EDI transmissions to identify a party to its trading partners.

CommerceNet A non-profit organization open to public and private organizations which has a charter to develop, maintain, and promote an Internet-based infrastructure for ELECTRONIC COMMERCE in business-to-business applications.

Commercial Internet Exchange Association (CIX) An association founded by a group of INTERNET SERVICE PROVIDERS to provide unrestricted BACKBONE services for business use and to use agreed-upon procedures in accounting for commercial traffic.

common carrier A regulated, private company that provides voice and/or data communication services to the public for a fee.

Common Gateway Interface (CGI) A standard for interfacing between some WORLD WIDE WEB servers and external GATEWAY programs written by users. Common uses of a CGI gateway include converting user documentation into HTML on the fly and sending the HTML result to a CLIENT, interfacing with WAIS and archive databases and converting the result into HTML to be sent to a client, and obtaining user feedback about a WEB SITE through an HTML form and a CGI decoder.

Communications Assistance for Law Enforcement Act of 1994
Also known as the FBI Wiretap Bill, this legislation mandates that the nation's telecommunications infrastructure facilitate government interception of voice and data communications. Opposed by EPIC and other civil libertarian groups, the bill was also opposed by telephone companies because of the cost of additional equipment required. The law allocated $500 million compensation to the telcos that has not been funded. Without the federal money the law may be unenforceable.

communications program A software program that enables a PC to communicate with other computers and with networks, including the Internet, over dialup telephone lines.

communicon Communication icon, sometimes called a smiley. A symbol constructed from punctuation and letters to communicate a facial expression or tone of voice. One communicon is :-) and signifies a smile.

compatibility The ability of products which perform different functions (e.g. software programs, network protocols, computers, and other products) to work with each other. See INTEROPERABILITY.

compliance program A procedure in which two or more EDI TRADING PARTNERS report periodically on their conformity to agreed upon standards for control and audit.

compression See DATA COMPRESSION.

Computer aided design (CAD) The design of physical objects (e.g. printed circuit boards) using workstations and sophisticated software to improve accuracy and productivity. CAD FILES can be sent from computer to computer over a network.

Computer Professionals for Social Responsibility (CPSR) An organization that studies and lobbies lawmakers on the social implications of computerization.

computer-based conferencing Conferencing where all communication is through computer I/O (e.g. keyboards and displays).

Conference attendees may or may not share a view of a common display.

computer-integrated manufacturing (CIM) The use of ONLINE shared information systems in manufacturing and resource planning to rapidly develop prototypes, improve quality control, and reduce waste.

conditional element A DATA ELEMENT which may or may not be included in an EDI SEGMENT depending upon the use of other data elements.

confidentiality The assurance that information is not disclosed or revealed to unauthorized persons.

Connectionless Broadband Data Service A European PACKET-switched multimegabit NETWORK service similar to SWITCHED MULTIMEGABIT DATA SERVICE.

Consumer Internet Those computers and users who can ACCESS and use interactive services on the Internet. All users and computers in the CONSUMER INTERNET are in the MATRIX.

convergence The hypothesis that the communications, information, and entertainment industries are moving together toward a common interactive BROADBAND media. Also see DIGITAL CONVERGENCE.

Cooperative Research and Development Agreement (CRADA) A format for cooperation between companies in a single industry and one or more offices of the federal government.

copyleft A kind of COPYRIGHT, common on the Internet but not yet recognized in law, that permits the distribution and copying of a work without charge, provided that no fees are charged and no modifications are made to the work.

copyright Federal law which gives authors and artists the exclusive right to publish their work or decide who else may publish it. Software and databases may be copyrighted to a limited extent. The 1976 copyright law requires that a copyrighted work subsist in original works of authorship fixed in any tangible medium of expression.

Core Internet Those computers and users which can provide Internet services such as FTP, TELNET, and the WORLD WIDE WEB. All users and computers in the CORE INTERNET are included in the CONSUMER INTERNET. See MATRIX.

corporate trade exchange (CTX) An ELECTRONIC FUNDS TRANSFER through the AUTOMATED CLEARING HOUSE system effected by use of a PAYMENT ORDER/REMITTANCE ADVICE EDI STANDARD TRANSACTION SET.

Corporation for Open Systems A consortium which develops test suites and tests products for INTEROPERABILITY.

COS See CORPORATION FOR OPEN SYSTEMS.

CPSR See COMPUTER PROFESSIONALS FOR SOCIAL RESPONSIBILITY.

cracker A person who attempts to gain ACCESS to computers for which he or she does not have AUTHORIZATION.

CRADA See COOPERATIVE RESEARCH AND DEVELOPMENT AGREEMENT.

credit card A uniquely numbered plastic card issued by a BANK or other company that permits rapid verification and issuance of credit over phone wires to enables the card's owner to buy goods and services.

credit-debit A NETWORK PAYMENT SYSTEM that sets up a USER account on a payment SERVER and allows the user to authorize charges against the account. The account may be set up to require a positive balance (the debit or check approach) or to allow charges to accumulate and be paid subsequently by the user (the credit approach). In contrast to DIGITAL CASH the CREDIT-DEBIT system can be readily audited. For examples see FIRST VIRTUAL INTERNET PAYMENT SYSTEM, NETBILL, and NETCHEQUE.

critical mass The concept that, because a network derives its value from the users on it, the growth and value of a network such as the Internet increases exponentially once a certain number of users participate.

CRL See CERTIFICATE REVOCATION LISTS.

crosspost To post the same MESSAGE to multiple Internet newsgroups. When inappropriate, it is called SPAMMING or carpet bombing.

cryptographic service message (CSM) A message that carries KEYS or related information controlling a generation of key distribution.

cryptography The art and science of making and keeping MESSAGES secure.

cryptoperiod The time during which a particular KEY is used. Short periods improve SECURITY because keys may be compromised over time and because the accumulation of large amounts of CIPHERTEXT using the same key facilitates unauthorized deciphering of the key.

CSM See CRYPTOGRAPHIC SERVICE MESSAGE.

CTX See CORPORATE TRADE EXCHANGE.

CUSDEC An EDIFACT standard customs declaration MESSAGE sent by an importer to customs detailing products being imported.

CUSRES An EDIFACT standard MESSAGE sent by a customs department in response to a CUSDEC message.

customer account analysis An EDI STANDARD TRANSACTION SET sent to a customer by a BANK to report account balances and transactions.

customer identifier In EDI, the code that identifies a customer organization.

customer initiated entry (CIE) A banking TRANSACTION performed at an ATM, a self-service banking terminal, a telephone key pad, or a PC. CIE provides customer convenience while reducing bank costs.

cyber- A prefix meaning a computerized version of something, most often implying a relationship to the Internet and to CYBERSPACE.

CyberCash Inc. A company which enables secure CREDIT CARD and DEBIT CARD transactions over the Internet by coordinating encrypted information between consumers and merchants and presenting the TRANSACTION to the MERCHANT BANKS in formats identical to those for conventional transactions. The company will also support DIGITAL CASH and MICROPAYMENTS.

cybercasting The delivery of current issues of periodicals and other text over the Internet.

cybermall See E-MALL.

cyber notary See DIGITAL NOTARY SYSTEM

cyberspace A term coined by William Gibson in his science fiction novel, *Neuromancer,* and now used to refer to all computers and forms of electronic communication as an environment that users can enter and experience, especially over the Internet.

cybrarian A term coined by M. Bauwens for corporate librarians who use external databases, E-MAIL, bulletin boards, and computer conferencing systems to provide services, as well as traditional books, periodicals, and databases.

cypherpunk A person who believes an individual's communications should be immune to surveillance, particularly by government agencies, and works toward that end by developing and distributing free ENCRYPTION programs.

. D .

daemon A program which runs constantly in the background and takes specified actions when it detects that a certain event or events have taken place (e.g. advising a USER when new E-MAIL has arrived).

DARPA Defense Advanced Research Projects Agency (now ARPA).

data compression The reduction of the number of BITS transferred without modifying the content of a MESSAGE and while permitting reconstitution of the original message by the receiving party. Data compression can decrease the amount of data to be transmitted by as much as 4:1 and can thereby substantially increase the effective data transfer rate of a MODEM or circuit. See MNP5 and V.42BIS.

data element In EDI, a data field which corresponds to a defined category of information (e.g. price). Each DATA ELEMENT has an identifying data element number, a name, a definition, a specified data type, and a minimum and maximum length. See SEGMENT.

data element separator In EDI, a character used to separate the elements in a SEGMENT.

Data Encryption Algorithm See DATA ENCRYPTION STANDARD.

Data Encryption Standard (DES) A U.S. government approved method of ENCRYPTION that uses 56-BIT PRIVATE KEY CRYPTOGRAPHY. DES, standardized as ANSI X3.92, is widely used by the financial industry where it is known as the DEA (DATA ENCRYPTION ALGORITHM). Export of DES out of the U.S. must be approved on a case by case basis and approval is rarely granted except for subsidiaries and overseas offices of U.S. companies. The export restriction is controversial because of substantial overseas demand for stronger encryption. Algorithms using PUBLIC KEY CRYPTOGRAPHY are significantly more secure than DES. See TRIPLE DES, RC2 AND RC4.

data integrity The assurance that a document or any other data, particularly in digital form, has not been altered or deleted in TRANSMISSION or in processing.

Data Interchange Standards Association (DISA) A non-profit organization that administers the X12 standard for the ANSI X12 subcommittee and provides news updates on EDI.

data separate from dollars In EDI, a process in which the remittance advice is sent directly from the payer to the payee while the funds are transferred separately by their respective BANKS.

data with dollars In EDI, an ELECTRONIC FUNDS TRANSFER process in which the remittance advice is passed from the payer to the payee through their respective BANKS, rather than directly between the companies.

data-jacking Interception and capture by an unauthorized party of data transmitted over a network.

database Information residing on one or more computers and organized so that specific information requested by a USER or CLIENT can be readily retrieved.

database marketing Tracking individual consumers and their past purchases in order to deliver customized marketing MESSAGES and offers. See ONE-TO-ONE MARKETING.

datagram In the TCP/IP PROTOCOL STACK, a SEGMENT for TRANSMISSION to which a network header has been added at the Internet layer. Datagram also sometimes refers to a segment. See FRAME.

DBS Direct Broadcast Satellite TV. A technology for sending audio, video, and data to many locations simultaneously via a satellite and small receivers.

DDA See DEMAND DEPOSIT ACCOUNT.

DEA See DATA ENCRYPTION STANDARD.

de facto standard A standard that is widely used by multiple manufacturers and software developers without having been created or approved by any standards body (such as ANSI or ITU-TSS). See DE JURE STANDARD.

de jure standard A standard which has been approved by a national or international standards body (such as ISO or ANSI). See DE FACTO STANDARD.

debit authorization An EDI STANDARD TRANSACTION SET sent by a buyer to its BANK or other FINANCIAL INSTITUTION approving in advance a debit of the buyer's account to pay for goods or services received.

debit card A uniquely numbered plastic card issued by a BANK or other company that permits immediate transfer of funds from a checking or savings account. Also known as an asset card. Unlike a CREDIT CARD, use of a DEBIT CARD requires a secret PERSONAL IDENTIFICATION NUMBER (PIN). One type of debit card is an ATM CARD. A second type is a national debit card, designed for payments at retail.

decryption The restoration of encrypted data to its original PLAIN TEXT or other readily usable state.

dedicated line A permanently connected telephone line for carrying voice or data between two organizations or two locations of the same organization. Dedicated data lines typically run at 56Kbps and their uses include communications between bank branches and a central computing facility, between larger companies and their EDI VANS, and between a company LAN and an INTERNET SERVICE PROVIDER.

demand deposit account (DDA) A BANK account, such as a checking account, which permits the payment or withdrawal of funds immediately upon demand by the account holder.

DES See DATA ENCRYPTION STANDARD.

DEX/UCS See DIRECT EXCHANGE OF UCS DATA.

dialup access ACCESS to an INTERNET SERVICE PROVIDER using a MODEM and a telephone line.

dialup port A PORT on an INTERNET SERVICE PROVIDER's computer that is connected to a MODEM and receives incoming data calls over the public telephone system from CLIENTS' computers.

digest Provided by the moderators of some newsgroups, a DIGEST groups postings so that users need not receive a multitude of individually forwarded E-MAIL MESSAGES. Some digests can be sorted by subject, sender, date, and THREAD.

DigiCash A company offering a proprietary method of enabling DIGITAL CASH using CRYPTOGRAPHY.

digital cash Funds held in an online account which can be transferred over the Internet between any two parties, including consumer to consumer. Digital cash may also be stored in ELECTRONIC PURSES and transferred using ELECTRONIC WALLETS. In both cases cryptography provides protection against theft. A customer purchases digital cash using a check, CREDIT CARD or other means. A major advantage of digital cash as a NETWORK PAYMENT SYSTEM is that it can be transferred anonymously (i.e. without the merchant knowing who the buyer is). It is especially useful for purchasing information for small dollar amounts (MICROPAYMENTS) directly over the network. Also see DIGICASH, and MONDEX.

digital certificate A MESSAGE which supports the use of a DIGITAL SIGNATURE by guaranteeing that the public KEY of the sender is indeed the property of the identified person. The certificate contains the users identifying information (name, organization, address, etc.), the users public key, the period during which the certificate is valid, the certificate serial number, and the digital signature and identifying information of the CERTIFICATE ISSUING AUTHORITY.

digital convergence The tenet that all forms of electronic communication - voice, data, cable and broadcast television, radio, cellular telephony, facsimile and others - will become digital, will increasingly share TRANSMISSION facilities, and will be presented in combination on a variety of intelligent devices. See

ASYNCHRONOUS TRANSFER MODE, CONVERGENCE, and MULTI-
MEDIA.

digital library The sum total of information available over the Internet - a very large library without a catalog system.

digital merchandise A FILE, document, or program offered and sold
in electronic form only.

Digital Notary System A system patented by Surety that uses CRYP-
TOGRAPHY to allow users to affix a time seal to the contents of any
computer record (e.g. a text, video, or image FILE). The seal can
be used after it is affixed to guarantee that the record existed in a
precise form at a precise time and has not subsequently been
altered. The system requires software on a CLIENT machine to
ACCESS the Surety Coordinating Server which returns a DIGITAL
CERTIFICATE for the specific record.

digital signature A method of assuring that a MESSAGE was sent by
the person claiming to send it. The signature is encrypted with
the sender's private key and decrypted by the recipient using the
sender's public key. Since only the sender could have encrypted
the signature the recipient is assured of the sender's IDENTITY.

Digital Signature Standard (DSS) A cryptographic technique for
authenticating electronic communications conforming to IEC
9796 International Digital Signature Standard and developed by
the National Security Agency (NSA) as part of the government's
CAPSTONE project.

Digital Telephony Act See COMMUNICATIONS ASSISTANCE FOR LAW
ENFORCEMENT ACT OF 1994.

digital watermarking A method of hiding the identification of the
original purchaser of a document within the document so that the
culpable party can be identified if a document is distributed
illegally.

direct access As opposed to DIALUP ACCESS, a HOST with DIRECT
ACCESS is on a network which is part of the Internet or on a LAN
which is connected to the Internet through a ROUTER. Without

an intermediate GATEWAY or computer between itself and the net, the CLIENT PC or workstation's ability to access and use resources on the Internet is constrained only by the client itself.

direct exchange of UCS data The manual delivery of EDI STANDARD TRANSACTION SETS in DIRECT STORE DELIVERY.

direct store delivery (DSD) Management of the replenishing of a retailer's stock by a manufacturer or distributor. See BLIND ENTRY and SMART CARD.

directory of servers On the Internet, a database that provides information about WAIS databases and supports searching of those databases.

DISA See DATA INTERCHANGE STANDARDS ASSOCIATION. (Note that DISA also stands for Defense Information Systems Agency.)

discount rate The percent of a BANK CARD TRANSACTION amount that an acquirer or MERCHANT BANK charges the merchant for giving the merchant credit and handling the TRANSACTION. Also see INTERCHANGE FEE.

distributed processing The use of multiple computers, connected over a network, to perform related data processing tasks.

DLL See DYNAMIC LINK LIBRARY.

DNS See DOMAIN NAME SYSTEM.

DOD Department of Defense.

domain An Internet entity with responsibility for naming groups and hosts that are subordinate to it.

domain name A word or group of characters that uniquely identifies a domain.

Domain Name Service The program on Internet routers and hosts which translates domain names into IP addresses.

domain name system (DNS) The method for naming Internet hosts without duplicating names.

download To transfer data from a SERVER to a CLIENT, usually over a network rather than over a direct connection.

DSD See DIRECT STORE DELIVERY.

DSD/UCS DIRECT STORE DELIVERY using EDI TRANSACTION SETS. See BLIND ENTRY.

DSS See DIGITAL SIGNATURE STANDARD.

dynamic link library (DLL) A FILE or files containing multiple PROTOCOL STACKS that can be selected automatically by APPLICATIONS.

· E ·

E-Form See ELECTRONIC FORM.

E-mail The delivery of text MESSAGES and attachments over a network from one computer to another, generally messages created by a human and intended for one or more other people. Basic Internet E-MAIL is limited to ASCII characters and does not permit files to be attached to messages. See MIME, UUENCODE.

E-Mall A WEB SITE maintaining electronic STOREFRONTS for multiple companies. The storefronts provide information and may be able to handle TRANSACTIONS.

e-money See DIGITAL CASH.

EAGLE An EDI standard used in the hardware and houseware industry.

EBB See ECONOMIC BULLETIN BOARD.

EBT See ELECTRONIC BENEFITS TRANSFER.

EC See ELECTRONIC COMMERCE.

EC/EDI See ELECTRONIC COMMERCE THROUGH ELECTRONIC DATA INTERCHANGE.

Economic Bulletin Board A Department of Commerce bulletin board service available on the Internet which provides over 2000 files of current economic information.

ECPA See ELECTRONIC COMMUNICATIONS PRIVACY ACT.

ECR See EFFICIENT CONSUMER RESPONSE.

EDA See ELECTRONIC DOCUMENT AUTHORIZATION.

EDI See ELECTRONIC DATA INTERCHANGE.

EDI Association An organization responsible for standardizing EDI TRANSACTION sets for the warehousing, transportation, and grocery industries.

EDI gateway A device through which EDI data is communicated externally to the organization. Also, the department within the organization that exchanges EDI data with the external business environment.

EDI standard transaction set An EDI TRANSACTION SET which conforms to the ANSI ASC X12 standard for EDI.

EDI system Software that links an EDI communication facility with business APPLICATIONS, translating between data formats used in the application and EDI TRANSACTION SETS.

EDI transaction set A TRANSACTION oriented document in a predefined digital format suitable for TRANSMISSION between company departments, separate companies, or FINANCIAL INSTITUTIONS. It usually contains a heading section, a detail section, and a summary section.

EDI VAN A value added network specifically designed to support EDI. Value added capabilities may include CARBON COPY ROUTING, conversion into FAX or mail, and automatic matching of line speeds and communication PROTOCOLS.

EDIA See EDI ASSOCIATION.

EDICA The Electronic Data Interchange Council of Australia. The Australian EDI administrative organization.

EDICC The Electronic Data Interchange Council of Canada. The Canadian EDI administrative organization.

EDIFACT United Nations-sponsored global EDI standards. EDIFACT stands for Electronic Data Interchange for Administration, Commerce, and Transport. EDIFACT is derived from the X12 standards but incorporates additional and different segments and uses a more flexible and generic approach to defining data elements using QUALIFIER CODE.

edu The top-level Internet DOMAIN NAME assigned to colleges and universities. In the United States, a college or university name always ends with a period followed by EDU (e.g. ucdavis.edu).

EDX The Electrical Industry Data Exchange, a group that develops EDI requirements for the electrical industry and works with the industry on the XI2 EDI standard.

EFF See ELECTRONIC FRONTIER FOUNDATION.

Efficient Consumer Response A grocery industry plan initiated in 1993 to improve the AVAILABILITY and promotion of goods to consumers by adopting EDI.

EFT See ELECTRONIC FUNDS TRANSFER.

EFT/EDI See FINANCIAL EDI.

EINet See ENTERPRISE INTEGRATION NETWORK.

electronic banking The operation of a BANK or the use of a bank using ELECTRONIC FUNDS TRANSFER.

electronic benefits transfer (EBT) The delivery of government benefits, such as public assistance and food stamps, through automatic teller machines and other electronic banking systems. EBT systems have been implemented in six states and most others are considering such a program. See REGULATION E.

Electronic Business Co-op A joint venture providing an end-to-end solution for buying and selling over the WEB using ENCRYPTION. The partners in the group are Spyglass (Web BROWSER), Tandem Computers (servers), V-One (ENCRYPTION), and CheckFree (PAYMENT SYSTEM).

electronic catalog 1) A catalog of products and prices, usually available on a CD ROM or over the Internet at a WEB SITE, which may or may not support ordering catalog items over the network. An electronic catalog can be searched quickly by the consumer, be updated as needed, and can present much more information about a product than the usual paper catalog. 2) In EDI, a catalog on a customers computer system which can be updated remotely.

electronic commerce (EC) the conducting of business communication and transactions over networks and through computers. As most restrictively defined, electronic commerce is the buying and selling of goods and services, and the transfer of funds, through digital communications. But EC also includes all inter-company and intra-company functions (such as marketing, finance, manufacturing, selling, and negotiation) that enable commerce and utilize E-MAIL, EDI, file transfer, FAX, VIDEO CONFERENCING, WORKFLOW or interaction with a remote computer (including use of the WORLD WIDE WEB). Also see NETWORK PAYMENT SYSTEMS and ELECTRONIC FUNDS TRANSFER.

Electronic Commerce Association A non-profit organization established to advance ELECTRONIC COMMERCE and business use of technology in Canada.

Electronic Commerce through Electronic Data Interchange (EC/EDI) A project of DOD and the Lawrence Livermore National Laboratory to build business systems for ELECTRONIC COMMERCE over the Internet using standard EDI formats.

Electronic Communications Privacy Act (ECPA) A U.S. law that makes unauthorized ACCESS to a computer system which transmits private communications a criminal offense. The law protects MESSAGES stored for up to 180 days but not longer.

electronic currency See DIGITAL CASH.

electronic data interchange (EDI) The exchange of standardized document forms between computer systems for business use.

electronic document authorization (EDA) A capability most common to WORKFLOW SOFTWARE systems which goes beyond AUTHENTICATION and DIGITAL SIGNATURE to provide assurance to the recipient of an ELECTRONIC FORM or MESSAGE that the sender has the authority or appropriate spending limit to sign and send the document.

electronic form A business form displayed and completed by a user on a computer screen rather than printed on paper. Using networks, electronic forms can be routed more quickly, efficiently, and accurately than traditional paper forms. See WORKFLOW SOFTWARE.

Electronic Frontier Foundation (EFF) An organization founded in 1990 to ensure that the principles embodied in the Constitution and the Bill of Rights are protected as new communications technologies emerge. EFF work focuses on new laws that protect citizen's rights in using new communications technologies and on common carriage requirements for all network providers so that all speech will be carried without discrimination.

electronic funds transfer (EFT) The transfer of funds by incrementing and decrementing accounts through electronic means without the use of checks or other paper. WIRE TRANSFER EFT generally refers to large dollar transactions among financial institutions and government entities, such as FED WIRE , AUTOMATED CLEARING HOUSE payroll direct deposit, business-to-business payment using EDI, and ELECTRONIC BENEFITS TRANSFERS. Consumer electronic PAYMENT SYSTEMS include AUTOMATED TELLER MACHINE , POINT-OF-SALE SYSTEMS using DEBIT CARDS, and bill payment by telephone or PC.

electronic mail A system for transferring information, primarily text MESSAGES generated by users, from one computer to another in ELECTRONIC FORM. ELECTRONIC MAIL is the most popular Internet APPLICATION and many users have ACCESS TO ELECTRONIC MAIL over the Internet without having access to any other Internet service.

electronic money See DIGITAL CASH.

electronic notary See DIGITAL NOTARY SYSTEM.

Electronic Privacy Information Center (EPIC) An organization which focuses public attention on the PRIVACY implications of the National Information Infrastructure, particularly CAPSTONE, the COMMUNICATIONS ASSISTANCE FOR LAW ENFORCEMENT ACT, medical record privacy, and the sale of consumer data.

electronic publishing The distribution over the Internet of newsletters, journals, white papers, research reports and periodicals through USENET groups, the WORLD WIDE WEB, or other means, regardless of whether or not the document was originally created for distribution over the network.

electronic purse A SMART CARD with stored value that can be loaded with DIGITAL CASH at banks or from ELECTRONIC WALLETS in the home and spent at properly equipped POINT-OF-SALE SYSTEM devices. It is designed to replace cash and coins for many consumer payments under $10. There are over 300 billion consumer cash transactions in the U.S. each year versus only 60 billion consumer BANK CARD, check, and wire transfer transactions. Visa International has formed a consortium of international leaders in consumer payments to develop common worldwide specifications for an electronic purse.

electronic signature The digital image of a persons signature, usually written on a handheld notepad, which can be transmitted. Also see DIGITAL SIGNATURE.

electronic storefront See STOREFRONT.

electronic wallet 1) A small portable device that loads and reads SMART CARDS. It can display how much money is stored on the user's own smart card, and transfer money (with the appropriate AUTHORIZATIONS) between the wallet and the user's or someone else's smart card Some electronic wallets may be able to communicate wirelessly and some may be integrated with PDAs. 2) As used by CheckFree Corporation and the ELECTRONIC BUSINESS CO-OP, a program which can be integrated into or used in conjunction with a WWW BROWSER to enable the secure TRANSMISSION of CREDIT CARD information over the network and permit CheckFree to authorize and settle the TRANSACTION for a merchant.

Electronics Industry Data Exchange A group that developed EDI data requirements for the electronics industry.

EMU See ENERGY MANAGEMENT UNIT.

encapsulation 1) Within a given level of a PROTOCOL STACK, the adding of a header to a block of data received from a higher level protocol to reformat the block before passing it to a lower level protocol or to the NETWORK for transmission. See INTERNET

PROTOCOL. 2) A method of transmitting data over dissimilar networks by enclosing the entire FRAME from one network in the header used by the other network.

Encapsulating Security Payload (ESP) An optional SECURITY header specified in the IPV6 protocol. See SECURITY PARAMETERS INDEX.

encryption The disguising of a MESSAGE to obscure its meaning. See also DECRYPTION.

energy management unit (EMU) A device that monitors and controls home appliances and thermostats to use electricity in off-peak hours, thereby saving money for the utility and the consumer. An EMU uses a TV screen or a PC for a display.

Enterprise Integration Network (EINet) A business network operated by MCC and designed to provide high-speed directory, ENCRYPTION, and ELECTRONIC FUNDS TRANSFER.

envelope See INTERCHANGE.

EPIC See ELECTRONIC PRIVACY INFORMATION CENTER.

Ethernet The most popular LOCAL AREA NETWORK technology, invented by Bob Metcalfe.

Eudora Popular E-MAIL client software from Qualcomm that handles ENCRYPTION and MULTIMEDIA.

Exchange of Product Model Data (STEP) An emerging international standard for describing product designs so that coordination between manufacturers can be facilitated.

ExpressNet A service from American Express that allows subscribers to review details of their American Express accounts, make travel arrangements, and correspond with other ExpressNet members.

· F ·

FAQ A list of Frequently Asked Questions about a specific topic with their answers. Often associated with USENET newsgroups and mailing lists.

FASA See FEDERAL ACQUISITION STREAMLINING ACT.

fax An abbreviation of facsimile, the electronic TRANSMISSION of a printed image or image formatted for printing.

FBOI See First Bank of the Internet.

FCC Federal Communications Commission. An agency of the U.S. Department of Commerce responsible for regulating broadcasting and telecommunications activities.

Fed Wire The Federal Reserve Wire Network. A high speed network operated by the U.S. Federal Reserve Bank to carry large dollar time-sensitive payments among federal agencies, most U.S. BANKS, and some other financial institutions. FED WIRE is also used by banks to transact business with the Federal Reserve Board, including borrowing from the Fed and buying and selling government securities. See also AUTOMATED CLEARING HOUSE.

Federal Acquisition Streamlining Act (FASA) A sweeping law enacted in October 1994, which furthers the use of ELECTRONIC COMMERCE and calls specifically for electronic notification of procurements and acceptance of bids.

FEDI See FINANCIAL EDI.

file A named set of information that can be stored or transmitted as a logical unit. A FILE may contain text, graphics, a program, or any other form of digital information.

File Transfer Protocol An Internet tool which allows users to transfer FILES from REMOTE HOSTS to their own computer.

filter A utility that disposes of unwanted input according to pre-defined specifications. In a ROUTER, a FILTER disposes of unwanted packets. In E-MAIL a filter automatically disposes of unwanted MESSAGES.

Financial EDI (FEDI or EFT/EDI) EDI transmissions between FINANCIAL INSTITUTIONS and companies which contain payment and remittance data such as bank statements, balance inquiries, deposit notices, PAYMENT ORDERS, and remittance advice.

financial information reporting An EDI STANDARD TRANSACTION SET reporting on account balances and transactions sent from a BANK to a customer.

financial institution A company or government organization that receives funds from consumers, businesses, or other institutions and invests the funds in financial assets, such as securities, loans, bank deposits, and property. FINANCIAL INSTITUTIONS include BANKS but the term is often used to refer to entities other than banks. Depository financial institutions accept deposits from the public which they channel into lending activities. They include banks, savings and loans, mutual savings banks and credit unions. Nondepositary financial institutions sell securities or insurance policies to the public rather than accepting deposits and include brokerage firms, life insurance companies, pension funds, and investment companies.

financial return notice An EDI STANDARD TRANSACTION SET sent to a customer by a FINANCIAL INSTITUTION that reports on items returned by an AUTOMATED CLEARING HOUSE.

Financial Services Technology Consortium (FSTC) A CRADA formed by several BANKS, universities, and federal laboratories to develop standards and technologies to support ONLINE banking and enhance the competitiveness of the U.S. financial services industry.

finger An Internet tool that allows a USER to obtain a list of users currently logged on to a specific HOST and find information on a specific user.

FIPS Federal Information Processing Standard.

fire wall A computer system, ROUTER or a pair of routers placed between the Internet and a private network to prevent unauthorized users from accessing the private network. A fire wall may utilize PACKET LEVEL FILTERING or an APPLICATION LEVEL GATEWAY and the most secure fire walls combine both.

First Bank of the Internet (FBOI) An Internet PAYMENT SYSTEM which uses E-MAIL and PGP encryption to carry checks from consumers to merchants and invoices from merchants to FBOI. The consumer's ATM card is registered with FBOI and allows FBOI to transfer funds from the consumers checking account to the merchant.

First Virtual Internet Payment System A unique CREDIT-DEBIT NETWORK PAYMENT SYSTEM for selling anything that can be distributed electronically. The system provides excellent SECURITY because neither BANK CARD information nor DIGITAL CASH is sent over the network. Buyers and sellers set up accounts with First Virtual. When a buyer downloads information the seller of the information informs First Virtual of the TRANSACTION and of the buyer's account identifier. First Virtual asks the buyer via E-MAIL to confirm purchase. Upon confirmation First Virtual debits the buyer's First Virtual account. On a regular basis First Virtual bills the buyer's CREDIT CARD for accumulated charges. Upon payment, First Virtual credits the seller's account. Buyers are able to obtain and review the information before committing to purchase it and the accounts of buyers who abuse this privilege are terminated.

flame An angry and often insulting MESSAGE, usually directed to the author of a USENET newsgroup POSTING.

flat-fee billing A method of pricing Internet ACCESS that provides unlimited use for a certain period of time, usually a month, in contrast with billing for the actual time used.

follow-up message A POSTING to a USENET newsgroup in reply to an earlier posting.

frame In the TCP/IP PROTOCOL STACK, a DATAGRAM to which a datalink header and a checksum has been added at the datalink layer. The physical layer of the stack converts the FRAME into electronic signals for TRANSMISSION.

Frame Relay A PROTOCOL for carrying unswitched data communication from a single HOST or ROUTER to a defined host or ROUTER through the public switched network.

free Internet access A service in which the INTERNET SERVICE PROVIDER does not charge for services but either requires subscribers to view a paid advertisement for a short length of time whenever they LOG ON to the service or receives a rebate from a long distance telephone company on the calls made to ACCESS the network.

free-net Internet ACCESS provided to schools, libraries, small organizations and members of the community at no charge.

freight details and invoice transaction An EDI STANDARD TRANSACTION SET providing the cost and related details of carrying a shipment.

FSTC See FINANCIAL SERVICES TECHNOLOGY CONSORTIUM.

FTP See FILE TRANSFER PROTOCOL.

functional acknowledgment (FA) An EDI STANDARD TRANSACTION SET that reports on receipt and validation of an EDI TRANSMISSION. See TRANSMISSION ACKNOWLEDGMENT and APPLICATION ACKNOWLEDGMENT.

functional group A group of EDI TRANSACTION sets of the same type sent together in a FILE that conforms to EDI standards.

functional group envelope The functional group header and trailer that identify the beginning and end of a FUNCTIONAL GROUP.

Fundamental Tenet of Cryptography If lots of smart people have failed to solve a problem then it probably wont be solved (soon). Stated by Kaufman, Perlman, and Speciner in Network Security, Private Communication in a Public World. The tenet is important because most popular ENCRYPTION methods cannot be proven to be secure. Mathematicians are constantly searching for ways to break CIPHERS quickly.

· G ·

GATEC See GOVERNMENT ACQUISITION THROUGH ELECTRONIC COMMERCE.

gateway A device that translates and manages communications between networks that are different in the protocols they use (e.g. frame relay to x.25) and in their overall design (e.g. the Internet and America Online).

GIF See GRAPHICS INTERCHANGE FORMAT.

GII See GLOBAL INFORMATION INFRASTRUCTURE.

giro A widespread European and Japanese electronic PAYMENT SYSTEM used by consumers in place of checks. A GIRO automatically transfers funds from the consumers account to a vendors account and notifies the vendor when the transfer is made.

Global Information Infrastructure (GII) A concept, derived from the National Information Infrastructure, which calls for a worldwide assembly of systems that integrate networks, computers, information, APPLICATIONS, and people to create new ways of learning, working, and interacting.

Global Network Navigator (GNN) An ONLINE guide to the Internet developed by OReilly & Associates that locates information on the Internet by category.

GNN See GLOBAL NETWORK NAVIGATOR.

Gopher An Internet tool for accessing information on GOPHER servers. See ARCHIE, VERONICA, and JUGHEAD.

Government Acquisition Through Electronic Commerce (GATEC) A government procurement program initiated by the Air Force and part of EC/EDI.

Graphics Interchange Format (GIF) A FILE format which supports the TRANSMISSION of graphics files through many systems and networks, including BULLETIN BOARD SYSTEMS and the Internet.

groupware Software and systems which help groups coordinate and communicate about work on which they are cooperating. Groupware may incorporate E-MAIL, shared databases, WORKFLOW SOFTWARE, conferencing software, and scheduling software.

· H ·

hacker A person who demonstrates technical knowledge and creativity on the network by achieving a difficult objective of little practical value or of substantial negative value (e.g. unauthorized ACCESS to a computer system, destruction of data, or the disruption of networks or systems). See CRACKER.

handshaking A well defined set of machine generated MESSAGES used to establish communication between two computers or other devices.

hardware independence The ability of software to operate on a variety of computers (e.g. PCs, UNIX workstations, Macintosh, and minicomputers) from different vendors.

hash function An algorithm that translates variable-sized input into a fixed-sized string (the hash value). If a HASH FUNCTION is one-way (difficult or impossible to invert) the result is called a MESSAGE DIGEST and can be regarded as the digital fingerprint of the original document. A message digest is valuable because it can uniquely and securely represent a much larger string. Hash functions are used to compute digital signatures and to assure that two copies of a message are identical without comparing either the PLAIN TEXT MESSAGE or an encrypted copy of it. They also enable digital time-stamping of a document without revealing the contents of the document to the time-stamping service. Well known hash functions include MD4, MD5, and SHS.

HDLC See HIGH-LEVEL DATA LINK CONTROL.

header record A record which may precede a group of EDI TRANSACTION SETS and contain control information.

High-Level Data Link Control (HDLC) An ISO standard data link layer PROTOCOL derived from SDLC which provides a method of encapsulating data on serial LINKS that use SYNCHRONOUS TRANSMISSION.

hit The accessing of a WEB page by a USER. The number of hits in a given period of time are used to measure the popularity of the page.

home automation The installation, use, and communication of sensors, processors, and energy management systems in home appliances, heating, and air conditioning.

home banking The use of home based PCs, INTERACTIVE TELEVISION, or the telephone (either by voice or by key pad entry) to pay bills or transact business that would be normally done at a bank teller's window. See CUSTOMER INITIATED ENTRY (CIE).

home office An office maintained in a household for a home based business, TELECOMMUTING, or working after hours.

home page The initial WEB PAGE through which a WEB SITE is usually accessed and which usually contains an introduction to the site and comprehensive LINKS to other pages at the site.

host Any computer which can run Internet protocols and which can act as either a CLIENT or a SERVER on the Internet.

HPCC High Performance Computing and Communications Program. A U.S. government multiagency project to support research on advanced networks, software, and supercomputers.

HTML See HYPERTEXT MARKUP LANGUAGE.

HTTP See HYPERTEXT TRANSFER PROTOCOL.

hub A large company which strongly urges their smaller TRADING PARTNERS, particularly vendors, to use EDI.

hypertext Documents, such as WEB PAGES, that contain LINKS to sections of the same document or to other documents and permit the user to JUMP to another location by selecting the appropriate link. Selection usually requires clicking on text which refers to the second location and appears in a different color from the main text.

HyperText Markup Language The document description language used for creating WORLD WIDE WEB pages.

▪ I ▪

I-Way Information highway or the CONVERGENCE of all information highways. Bob Metcalfe identified six I-Ways: corporate networks, video telephony, INTERACTIVE TELEVISION, BULLETIN BOARD SYSTEMS, ON-LINE SERVICES, and the INTERNET.

IAB See INTERNET ARCHITECTURE BOARD.

IANA See INTERNET ASSIGNED NUMBER AUTHORITY.

ICMP See INTERNET CONTROL MESSAGE PROTOCOL.

identity The human associated with a login name. See AUTHENTICATION.

IDTV Interactive Digital TV. An IDTV SET-TOP BOX performs digital decompression and decoding while handling user interaction.

IEEE The Institute of Electrical and Electronic Engineers.

IESG See INTERNET ENGINEERING STEERING GROUP.

IETF See INTERNET ENGINEERING TASK FORCE.

indirect access Access to the Internet through a GATEWAY provided by an ONLINE SERVICE or an INTERNET SERVICE PROVIDER, limiting the user to Internet services offered by the gateway provider.

Infobahn The confluence of information services and networks, especially of high speed transmissions and video technology. See DIGITAL CONVERGENCE.

Infohaus An implementation of the FIRST VIRTUAL INTERNET PAYMENT SYSTEM which handles billing, collections, and payment for its sellers automatically.

information appliance A computer, viewed as an everyday commodity product.

information float The length of time between when a MESSAGE or other information is sent and the time it is received. For most

E-MAIL messages on the Internet, the Information float is only a few minutes.

information furnace A theoretical device that processes information for every video screen, personal computer, and telephone in a household.

information metering See METERING.

information security The pursuit of CONFIDENTIALITY (or PRIVACY), DATA INTEGRITY, AVAILABILITY, and LEGITIMATE USE of information in a computer system or NETWORK.

information superhighway The GLOBAL INFORMATION INFRASTRUCTURE. Also see I-WAY, INFOBAHN, and DIGITAL CONVERGENCE.

Information Technology (IT) The department in an organization that manages the acquisition and maintenance of computers and the development of software and SYSTEMS central to the organization's function. IT may or may not be responsible for departmental computers and LANs, for WANs and for telephony. The term IT has generally superseded MIS (Management Information Services).

Information Vending Encryption System (IVES) An AT&T SECURITY system for the marketing, sale, and distribution of information and information entitlements (including video-on-demand, electronic publishing, software distribution, home shopping, program execution monitoring, and home banking services) over data networks, BROADBAND MULTIMEDIA networks, and direct broadcast satellite TV (DBS). The system uses chips jointly designed by AT&T and VLSI.

I-Net The Internet.

Insurance Value Added Network Services (IVANS) An organization that links property/casualty insurance companies and agents to provide insurance-related electronic communications.

Integrated Services Digital Network (ISDN) Protocols developed by telephone companies worldwide to support digital TRANSMIS-

45

SION and higher data rates than possible with conventional analog technology and to improve control over voice and data circuit switching. See BASIC RATE INTERFACE and PRIMARY RATE INTERFACE.

integrated services network A network which can seamlessly handle all types of asynchronous and synchronous data including data, voice, and video. Real world networks of this type are expected to be based upon CELL RELAY and ATM technology.

integrity Assurance of the consistency of data, particularly the prevention of unauthorized creation, modification, or destruction of data.

intellectual property Documents, artwork, databases, or software which may or may not be protected under COPYRIGHT.

interactive television A technology which allows a television viewer to communicate with the content provider to select on-screen options and, in some cases, order products and services for delivery.

interchange 1) In EDI, all the data sent from one party to another in a single TRANSMISSION. Such a transmission begins and ends with an interchange header and interchange trailer which form an interchange envelope. 2) The name of a commercial online service operated by AT&T.

interchange control structure The basic structure of an EDI INTERCHANGE consisting of (in the X12 standard) three levels of ENVELOPE: the INTERCHANGE ENVELOPE, the FUNCTIONAL GROUP ENVELOPE, and the TRANSACTION SET envelope.

interchange envelope See INTERCHANGE.

interchange fee In a BANK CARD TRANSACTION, the amount paid by the ACQUIRER (MERCHANT BANK) to the ISSUER for the use of funds prior to SETTLEMENT and other operating costs. Interchange fees are set by MasterCard and Visa.

intermarketing The use of computers and the network to flexibly respond to customer inquiries and obtain customer input. See ONE-TO-ONE MARKETING.

internal order An order for material or work sent within a company, sometimes using EDI or resulting from an external order received through EDI.

International Standards Organization (ISO) The international standards organization of which ANSI is a member. ISO has defined the ISO/OSI protocols which are technically superior to TCP/IP but which are seldom used because of TCP/IP's widespread use and momentum. ISO has also defined standards for identification cards and financial TRANSACTION cards (see PAYMENT CARD).

internet Internetworking, the connecting together of multiple networks.

Internet The world-wide NETWORK of networks which provides ELECTRONIC MAIL, news, remote login, file transfer, and other services. Also see WORLD WIDE WEB, GOPHER, FILE TRANSFER PROTOCOL, and TELNET.

Internet Architecture Board (IAB) A council appointed by the INTERNET SOCIETY that holds the ultimate authority for standards and other Internet related issues. In practice, the IAB generally approves what the INTERNET ENGINEERING TASK FORCE proposes.

Internet Assigned Number Authority (IANA) An organization supervised by the IAB which controls all the network address numbers in the TCP/IP PROTOCOL suite. The INTERNET NIC and the Domain Name Administration report to IANA.

Internet Control Message Protocol An IP layer protocol which reports problems in the TRANSMISSION of datagrams, requests that hosts temporarily stop transmitting, requests time stamps and other information, and allows the network to adapt to congestion and other disruptions.

Internet dial tone Access to Internet BANDWIDTH, viewed as a commodity.

Internet Engineering Steering Group (IESG) The part of the Internet Architecture Board which oversees the INTERNET ENGINEERING TASK FORCE.

Internet Engineering Task Force (IETF) A group that develops Internet standards. The standards are considered by the INTERNET ENGINEERING STEERING GROUP with appeal to the INTERNET ARCHITECTURE BOARD.

Internet PCA Registration Authority (IPRA) The Level 1 organization in the PEM model which provides certification to multiple POLICY CERTIFICATION AUTHORITIES.

Internet presence A company's provision of a means for Internet subscribers to ACCESS at least minimal information about the company. The most common means are a WWW HOME PAGE and E-MAIL with automated reply.

Internet Protocol The PROTOCOL for the internet layer (layer three) of the TCP/IP PROTOCOL STACK. This protocol is responsible for addressing and sending blocks of data across the network. It adds network headers to segments received from the transport layer above it (layer four) in order to form DATAGRAMS. The datalink layer, below the internet layer, adds a datalink header and trailer and a checksum to convert the datagram into a FRAME for transmission over the network. See PACKET, TRANSMISSION CONTROL PROTOCOL/INTERNET PROTOCOL, and INTERNET CONTROL MESSAGE PROTOCOL.

Internet Protocol Security Protocol (IPSP) Optional additions to the IP PROTOCOL, being developed by IETF, to facilitate ENCRYPTION over the Internet.

Internet Research Steering Group (IRSG) The part of the INTERNET SOCIETY which coordinates the INTERNET RESEARCH TASK FORCE (IRTF).

Internet Research Task Force (IRTF) A group which addresses long-term INTERNET ISSUES.

Internet service provider (ISP) A company or organization that provides connections to the Internet. There are over three hundred ISPs worldwide.

Internet Shopping Mall A USENET list of businesses which can be contacted over the Internet.

Internet Society (ISOC) A voluntary membership organization that leads technological development of the INTERNET. ISOC appoints the INTERNET ARCHITECTURE BOARD.

InterNIC An organization funded by the NATIONAL SCIENCE FOUNDATION, which provides registration, database, and information services to users of the Internet world wide.

Interop See NETWORLD + INTEROP.

interoperability The ability to integrate products from different vendors into functional SYSTEMS without developing custom software, hardware or tools, and without using GATEWAYS between the products. Increasing levels of interoperability can be defined as follows: Coexistence means products can both work in the same system without interference but without cooperation. Solution-specific interoperability limits the joint use of products to a specific purpose and may require extensive custom engineering and configuration. Compatibility means parts can fit together and may be able to work together. Plug and play interoperability allows parts and products of any of the manufacturers to be plugged into the system and easily integrated. The highest level of interoperability is interchangeability, where products from different manufacturers are functionally identical, like light bulbs, and can be used interchangeably.

inventory availability inquiry An EDI transaction set that requests information on the availability of a specified product or products. Availability is most commonly expressed in minimum days required between receiving an order and shipping the product.

IP See INTERNET PROTOCOL.

IP address A 32-BIT number that uniquely identifies the connection between the Internet and each HOST computer or IP ROUTER. See DOMAIN NAME SYSTEM.

IP spoofing A method of gaining unauthorized ACCESS to a computer or network by constructing and sending packets that appear to come from a trusted computer, often a computer that is on the network being attacked. Access through IP SPOOFING can be prevented with an APPLICATION LEVEL GATEWAY FIRE WALL.

IPng IP next generation. A recommendation of the IETF for a new IP PROTOCOL which was approved by the IESG in November 1994. The new protocol is named IPV6.

IPRA See INTERNET PCA REGISTRATION AUTHORITY.

IPSP See INTERNET PROTOCOL SECURITY PROTOCOL.

IPv6 A new version of the IP PROTOCOL which will interoperate with the current version (IPv4). IPV6 will support a very large increase in the number of available IP addresses using 128-BIT hierarchical addresses, provide a flow control label for identifying real time traffic, provide flexible header extension for SECURITY (separate headers for AUTHENTICATION and for PRIVACY) and route selection, standardize MULTICAST, and provide plug and play autoconfiguration. First software implementations are expected in late 1995. See IPNG and SECURITY PARAMETERS INDEX.

IRC Internet Relay Chat An Internet tool that allows users to exchange MESSAGES interactively and in real time.

IRSG See INTERNET RESEARCH STEERING GROUP.

IRTF See INTERNET RESEARCH TASK FORCE.

ISDN See INTEGRATED SERVICES DIGITAL NETWORK.

ISO See INTERNATIONAL STANDARDS ORGANIZATION.

ISOC See INTERNET SOCIETY.

ISP See INTERNET SERVICE PROVIDER.

issuer A FINANCIAL INSTITUTION which issues a BANK CARD and authorizes TRANSACTIONS made with the card.

IT See INFORMATION TECHNOLOGY.

item authorization transaction An EDI STANDARD TRANSACTION SET used in grocery industry DIRECT STORE DELIVERY that specifies the delivery schedule and terms of sale of a particular product.

item information transaction An EDI STANDARD TRANSACTION SET that supplies product information such as PRODUCT CODE, price, size, and product description.

ITU-TSS International Telecommunication Union Telecommunication Standardization Sector, the standard setting body of the IEEE. The ITU-TSS has taken over virtually all of the responsibilities of the former CCITT.

ITV See INTERACTIVE TELEVISION.

IVANS See INSURANCE VALUE ADDED NETWORK SERVICES.

IVDS Interactive Video and Data Services using 218 to 219 megahertz set aside by the FCC for interactive TV over short distances.

IVES See INFORMATION VENDING ENCRYPTION SYSTEM.

▪ J ▪

JIT See JUST-IN-TIME.

JPEG Joint Photographic Experts Group standards for compression of images. JPEG attains much greater compression than GIF by selectively losing graphic information which is not normally detectable by the human eye.

Jughead A CLIENT APPLICATION which searches for directory titles on worldwide GOPHER servers. See VERONICA.

jump To move from one HYPERTEXT location to another by selecting an ANCHOR or LINK.

Just-in-time manufacturing schedule A schedule that minimizes inventory buffers between sequential manufacturing processes.

Just-in-time (JIT) A management approach that minimizes inventory and drives the improvement of the manufacturing process by requiring that components and raw materials arrive in small lots just as they are needed and that products be manufactured shortly before they are due to be shipped. JIT requires close coordination between companies and can make excellent use of robust EDI relationships.

· K ·

Kbps Kilobits per second. Thousands of BITS PER SECOND data transfer rate.

KDC See KEY DISTRIBUTION CENTER.

Kerberos An AUTHENTICATION system developed at MIT designed to authenticate requests for network resources rather than to authenticate originators of documents. The system uses a Kerberos SERVER on the network to store the private KEYS of all users and verify their IDENTITY upon request. Upon verification, the server issues a SESSION KEY for use between the user and the requested network RESOURCE.

key A number used to encrypt or decrypt data. Generally the longer the number the stronger the key and the harder it is to break CIPHERTEXT encoded with it. DES uses a 56 BIT key while CAPSTONE uses an 80 bit key. RSA commonly uses 512 bits. RC2 AND RC4 use variable length keys. See PRIVATE KEY and PUBLIC KEY.

Key Distribution Center (KDC) A party on a network that is trusted by two or more entities and which issues SESSION KEYS for their use in communicating with one another upon request.

key escrow The keeping of a cryptographic KEY by a trusted party for recovery and use in well defined circumstances (e.g. the departure of the employee who used the key). Also see LAW ENFORCEMENT ACCESS FIELD.

key management The development and control of public and private KEYS within an organization. Key management includes key generation, distribution, activation and deactivation, replacement, revocation, and termination (involving key destruction and possibly archiving). See also KEY ESCROW.

key recovery See KEY ESCROW.

kill file A FILEcreated by a USENET user that excludes postings coming from certain individuals, or containing certain words or subjects. See MAIL BOMB.

knowbot A computer program that runs independently once activated by a USER. A KNOWBOT finds specified information and delivers it to the user who requested it. Current knowbots, such as the San Jose Mercury's NewsHound, Compuserve's Executive News Service, and the DARPA Electronic Library Project at Stanford (elib@db.stanford.edu) search limited sets of sources. Future knowbots are expected to search virtually the entire Internet with more advanced algorithms.

▪ L ▪

LAN See LOCAL AREA NETWORK.

Law Enforcement Access Field (LEAF) Data included in MESSAGES encrypted by the CLIPPER chip which enables government agencies to obtain the serial number of the sending chip and an encrypted version of the key used to encrypt messages in a given session. Under court order the law enforcement agency can obtain the two parts of the key which is unique to the sending chip and which can be combined to decrypt the SESSION KEY and then use the session key to decrypt the message.

LEAF See LAW ENFORCEMENT ACCESS FIELD.

leased line See DEDICATED LINE.

legitimate use The assurance that users of a computer system or network are limited to utilizing resources only in authorized ways.

link 1) A logical connection between one HYPERTEXT page or document and another that permits a user to JUMP directly from the first location to the other. 2) A circuit or TRANSMISSION path between a sender and a receiver.

linked content The information contained in the WORLD WIDE WEB.

LISTSERV A BITNET MAILING LIST utility that automatically redistributes MESSAGES sent to it to all members of the MAILING LIST.

local area network A data network serving a small area (generally a floor or a small building) with relatively high data rates and low error rates. See WIDE AREA NETWORK.

lock box A secure location, such as a post office box, convenient to a customer for depositing payments, where a local BANK receives the payments and transfers them to the vendors bank in order to speed receipt of cash.

lock box transaction An EDI STANDARD TRANSACTION SET that details payments into a LOCK BOX and the balance of funds in the lock box.

log on A procedure for gaining ACCESS to a computer system, AP-PLICATION, or other RESOURCE that usually requires the input of a user name and a PASSWORD.

log-on ID A user name, unique to the system, which identifies the user to the system.

look-to-buy ratio The ratio between the number of times a web page is viewed to the number of times an item on the page is purchased, either over the network or as a result of viewing the page.

look-to-request ratio The ratio between the number of times a web page is viewed to the number of times a free item offered on the page, such as a document, catalog or brochure, is requested for delivery.

looping In a single EDI TRANSACTION SET, the repetition of a set of segments with different information in each repetition of the set. Line items in an invoice are an example.

· M ·

MAC Macro authentication code. A sequence of characters calculated from the contents of an EDI transmission. The MAC is added to the end of the transmission and compared to the receiver's independent calculation of the MAC to authenticate the transmission. See AUTHENTICATION.

mail bomb To fill up another Internet users E-MAIL MAILBOX with large unwanted messages. See KILL FILE.

mail float The length of time between when a check is written and when it is presented to a BANK for payment.

mailbox 1) A logical storage area for E-MAIL MESSAGES that are incoming to a USER or which have been retained by a user. 2) In EDI, a logical storage area where all incoming EDI data is kept until retrieved by the receiving party, usually by dial-up access. EDI VANs provide mailboxes for their customers.

mailing list A discussion group that uses ELECTRONIC MAIL to communicate.

MAN metropolitan area network.

mandatory data element or segment A DATA ELEMENT or SEGMENT in an EDI STANDARD TRANSACTION SET in which information must be contained.

manufacturing resource planning (MRP) A discipline which aims to assure completion of finished quantities at the time they are due. MANUFACTURING RESOURCE PLANNING (MRP) has largely incorporated and superseded materials requirements planning (also known as MRP).

Manufacturing Technology Centers (MTC) Seven regional offices administered by NIST designed to assist small and medium sized businesses with technology analysis, information, and ACCESS to training, management, financial, and marketing services.

map In EDI, the translation of EDI data from one EDI format to another or between an EDI format and the format of another system.

MAP Manufacturing Automation Protocol

MarketNet A company in the UK which provides services for selling over the Internet using SSL.

master key A key which is used to encrypt other keys (PRIMARY KEYS) rather than to encrypt data. Also known as a key-encrypting key.

material release An EDI TRANSACTION SET which authorizes and requests delivery of product which has previously been ordered.

material requirements planning See MANUFACTURING RESOURCE PLANNING.

Matrix As defined by John Quartermain and Smoot Carl-Mitchell of MDIS, the set of INTERNET users who can exchange E-MAIL with each other. They estimate that as of October 1994 there were 27.5 million users in the MATRIX, 13.5 million users and 3.5 million computers in the CONSUMER INTERNET , and 7.8 million users and 2.5 million computers in the CORE INTERNET.

MBONE The Multicast backBONE of the Internet.

Mbps Megabits per second. Millions of BITS PER SECOND data transfer rate.

MCC Microelectronics and Computer Technology Corp., a government/industry consortium developing the ENTERPRISE INTEGRATION NETWORK.

merchant bank See ACQUIRER.

message 1) From the point of view of a human USER or a HOST computer, information sent and received, optionally including the addresses of the sender and receiver, in a format intelligible to the human or computer, but exclusive of all headers, checksums, or other control characters needed for TRANSMISSION. 2) In the TCP/IP PROTOCOL STACK, user data for transmission to which an APPLICATION header has been added at the application layer. See

SEGMENT. 3) In the EDIFACT standard for EDI, an electronic document.

metering 1) The measuring and reporting, usually remotely, on the use of software or pieces of information that is distributed in encrypted form via CD-ROM or by data broadcast over satellite. Software and special hardware in the end user's PC decrypts the product and reports the usage to a central site for billing. Metering is expected to give small companies ACCESS to databases and software which they could not purchase outright. See PAY-PER-USE and MICROPAYMENT. 2) The tracking of the number of authorized users under a software site license.

MFJ See MODIFIED FINAL JUDGMENT.

micropayment A very small payment (fractions of a cent to several dollars) over the Internet. See DIGITAL CASH.

Microsoft Network A forthcoming online service which will include access to the Internet using MICROSOFT WINDOWS 95 as a user interface.

mil The top-level Internet DOMAIN NAME assigned to U.S. government military organizations. A military domain name always ends with a period followed by MIL (eg. USAF.mil).

MIME See MULTIPURPOSE INTERNET MAIL EXTENSIONS.

MIME Object Security Service (MOSS) An improvement over PRIVACY ENHANCED MAIL proposed by David Crocker which requires a formal global hierarchy for certifying KEYS.

Minitel A data network run over the French telephone system with terminals in many businesses and most French homes. Minitel originated in the early 1980s as a directory service to replace paper phone books but is now used for many other APPLICATIONS, including making reservations for transportation and public events.

MIPS year One million instructions per second of computing power running for one year. MIPS years are used to measure how long it would take fast computers to crack the best commercially available ENCRYPTION schemes (on the order of 70 million MIPS years).

MMDS See MULTICHANNEL MULTIPOINT DISTRIBUTION SYSTEM.

MNP5 Microcom Networking Protocol, Class Five. A lossless DATA COMPRESSION PROTOCOL which has a maximum compression factor of 2:1 and is an alternative to V.42BIS.

modem (modulator/demodulator) A device that connects a digital computer or terminal to an analog TRANSMISSION line, typically a telephone line. Modems are used for both dialup and dedicated line ACCESS to the Internet. Common MODEM speeds (in BITS PER SECOND) and the corresponding standards are: 1200bps (V.22), 2400 (V.22bis), 9600 (V.32), 14,400 (V.32 bis) and 28,800 (V.34). Also see DATA COMPRESSION.

moderator A person who volunteers to screen and forward postings to USENET and MAILING LIST discussion groups.

Modified Final Judgment (MFJ) A consent degree between AT&T and the Justice Department which broke up AT&T at the beginning of 1984 into a reorganized AT&T and seven regional Bell holding companies.

Mondex A DIGITAL CASH system being tested in the UK and supported by NatWest Bank and Midland Bank. The system uses SMART CARDS as ELECTRONIC PURSES and allows DIGITAL CASH to be transferred using an ELECTRONIC WALLET, cash dispensers (AUTOMATED TELLER MACHINES), pay phones, and special Mondex home phones. The Mondex card can carry up to five currencies simultaneously and can be locked with a personal code so that only the cards owner can use the value on it.

Mosaic A BROWSER for the WORLD WIDE WEB that was developed by NCSA at the University of Illinois at Champaign-Urbana, available in both commercial and unsupported freeware versions.

MOSS See MIME OBJECT SECURITY SERVICE.

MPEG The Motion Picture Experts Group and the standard it has developed for digitizing and compressing video into files which can be transmitted over networks. MPEG is optimized for TRANS-MISSION speeds of one half to 1.5 megabits per second. Current versions are MPEG1, optimized for CD ROM and sub-T1 data rates and MPEG2 for DBS, cable, and broadcast applications.

MRP See MANUFACTURING RESOURCE PLANNING.

MRP-II Manufacturing resource planning which allows each work center to make the scheduling decisions for its own work.

MTC See MANUFACTURING TECHNOLOGY CENTERS.

multicast The sending of the same packets to multiple specific network addresses, as opposed to just one address or broadcast to all network addresses.

Multichannel Multipoint Distribution System (MMDS) A wireless cable technology which enables BROADBAND video and data transmissions at a fraction of the investment for wire and cable.

multihosting The renting out of space on a WEB SERVER to provide customized WEB sites to multiple organizations. Unlike an E-MALL, each organization's HOME PAGE is accessed directly by users, rather than by going through the E-mall's home page.

multimedia Documents that combine multiple data types, for example: text, graphics, scanned images, audio, and video.

multipoint-to-multipoint A mode of communication in which each party communicates with many other parties, as in electronic bulletin boards.

Multipurpose Internet Mail Extensions (MIME) A standard for sending MULTIMEDIA non-ASCII MESSAGES over the Internet. Basic Internet E-MAIL, conforming to the SMTP standard, is limited to ASCII text. MIME can be used to transmit digital video, GIF graphics, digitized sound, PostScript formatting, and enhanced character sets without modification. A new standard being worked on by the IETF will allow EDI transactions and INTERCHANGES in MIME.

· N ·

NACHA See NATIONAL AUTOMATED CLEARING HOUSE ASSOCIATION.

National Automated Clearing House Association (NACHA) T h e association of BANKS and other financial institutions through which members transfer funds from one to another electronically. See AUTOMATED CLEARING HOUSE.

National Information Infrastructure Testbed Coalition An organization that researches high-speed networks.

National Information Infrastructure (NII) The title of a STATE-MENT published by the Clinton/Gore administration in September 1993 promoting a seamless web of communications networks, computers, databases, and consumer electronics that will put vast amounts of information at users fingertips....[That] can help unleash an information revolution that will change forever the way people live, work, and interact with each other. The National Information Infrastructure is often referred to a the INFORMATION SUPERHIGHWAY. Also see XIWT.

National Initiative for Product Data Exchange (NIPDE) A government-facilitated but industry-led partnership that coordinates product standardization, implementation, and education efforts within and across industries. See EXCHANGE OF PRODUCT MODEL DATA.

National Institute of Standards and Technology (NIST) An office of the Department of Commerce which maintains seven regional MANUFACTURING TECHNOLOGY CENTERS and administers the AD-VANCED TECHNOLOGIES PROGRAM.

National Research and Education Network (NREN) A high performance network intended to grow out of the current INTERNET, serve research and educational institutions, and be a catalyst for the National Information Infrastructure.

National Science Foundation (NSF) A U.S. government agency that sponsors scientific and engineering research and education. NSF administered an important INTERNET BACKBONE, NSFNET. NSFNET ceased operation in 1995 because the private sector then provided sufficient backbone service. NSF has begun work on the NATIONAL RESEARCH AND EDUCATIONAL NETWORK.

NetBill A project of the Information Networking Institute at Carnegie Mellon University, backed by Visa, to design NETWORK PAYMENT SYSTEM protocols and software for charging for information and services delivered over the Internet. NETBILL will provide AUTHENTICATION, account management, TRANSACTION processing, and billing and reporting services for clients and users. NetBill is optimized for MICROPAYMENTS (e.g. 1 cent transaction cost for a 10 cent purchase) and for certifying delivery of the goods or services over the network. The system sends an encrypted copy of the information being purchased to the customer and then sends the key needed for DECRYPTION when payment is received.

NetCheque project A CREDIT-DEBIT NETWORK PAYMENT SYSTEM, developed by the Information Sciences Institute at the University of Southern California, that uses distributed accounting servers to support writing and accepting digital checks in very much the same manner as paper checks. The check writer obtains a ticket or proxy from a KERBEROS server that authenticates the user to an accounting server. The accounting server will deposit the check in the payee's account if the payee is on the same server or relay the check to the payee's accounting server.

NetChex A service that provides secure check writing capability over the Internet using an extension of an individual's existing checking account.

NetComm A demonstration by Southern California Edison using radios on utility poles to relay signals from a neighborhood to a central office, and POWER LINE CARRIER to connect the radio to SMART METERS at nearby houses. See CELLNET.

netiquette Considerate and responsible use of the Internet resources, particularly USENET. Some top level guidelines include considering the needs and egos of others, not generating unnecessary traffic, voluntarily limiting commercial use of the Internet, and not over reacting to breaches of netiquette by others.

NetMarket The first company to offer secure transactions over the World Wide Web, initially using PRETTY GOOD PRIVACY. NetMarket provides customized marketing and sales services to companies doing business over the Internet. In November 1994 NetMarket was acquired by CUC International, a multibillion dollar company that provides membership services such as home shopping.

Netscape A company founded by Marc Andreesen, co-developer of MOSAIC and Jim Clark, founder of Silicon Graphics, Inc. The company is a leader in WWW browsers and servers.

Netscape Navigator A product of Netscape Communications Corporation and the most popular WWW BROWSER.

network 1) The hardware and software that permits multiple computers to communicate with one another electronically or using fiber optics. 2) The INTERNET.

Network File System (NFS) A PROTOCOL suite and software developed by Sun Microsystems that allows ACCESS to files over a network and manipulation of the files as though they were local.

Network Information Center (NIC) Any one of several organizations that provide information about all of the INTERNET or a portion of it for a particular audience. See INTERNIC.

network payment system A PAYMENT SYSTEM which operates over an open unsecure network such as the Internet. The three basic models for NETWORK PAYMENT SYSTEMS are ELECTRONIC CURRENCY (e.g. DIGITAL CASH), CREDIT-DEBIT instruments (e.g. FIRST VIRTUAL INTERNET PAYMENT SYSTEM, NETBILL, and NETCHEQUE), and secure CREDIT CARD transactions (as supported by OPEN MARKET and CYBERCASH).

Networld + Interop The largest and most important network-related trade show, held twice a year in the United States and also held overseas.

network security See INFORMATION SECURITY.

newbie A new Internet USER.

NEX/UCS The EDI STANDARD TRANSACTION SETS used by the grocery industry in DIRECT STORE DELIVERY.

NFS See NETWORK FILE SYSTEM.

NIC See NETWORK INFORMATION CENTER.

NII See NATIONAL INFORMATION INFRASTRUCTURE.

NIPDE See NATIONAL INITIATIVE FOR PRODUCT DATA EXCHANGE.

NIST See NATIONAL INSTITUTE OF STANDARDS AND TECHNOLOGY.

non-repudiation Assurance to either party in an electronic TRANSACTION that the other cannot later deny having agreed to the transaction. The assurance is provided by means of a DIGITAL SIGNATURE.

NREN See NATIONAL RESEARCH AND EDUCATION NETWORK.

NSF See NATIONAL SCIENCE FOUNDATION.

NTIA National Information and Telecommunications Administration.

NTSC National Television Standards Committee.

· O ·

OBDC See OPEN DATABASE CONNECTIVITY.

ODETTE An EDI standard of the automobile industry in Europe.

Office of Technology Assessment (OTA) An office of the U.S. Congress that investigates and reports on many of the technical, regulatory, and economic issues raised by advancing communication and information technologies.

ONA See OPEN NETWORK ARCHITECTURE.

one-to-one marketing A marketing philosophy espoused for the future by Don Peppers and Martha Rogers which focuses on long-term customer retention through the use of two-way, individually addressable media.

online A computer or USER is online when it is connected to a NETWORK or a SERVER and is capable of receiving signals from the server or another computer.

online service A company providing news, E-MAIL, selected databases, and other services directly to consumers and business over dialup and other ACCESS methods. Examples include America Online, Prodigy, and CompuServe.

onramp An INTERNET SERVICE PROVIDER.

Open Database Connectivity (OBDC) A set of drivers originally defined by Microsoft that allow any database management system or front-end tool that supports OBDC to interoperate with any other that also uses OBDC.

Open Electronic Commerce The use of EDI STANDARD TRANSACTION SETS exchanged between Internet and VAN customers within standard E-MAIL envelopes using MIME and PEM enhancements. An Air Force pilot project with 1,800 bidders and suppliers demonstrated feasibility, cost savings, faster delivery, and improved productivity.

Open Market (OMI) A company that develops and markets a variety of software, services, and custom solutions to facilitate ELECTRONIC COMMERCE on the Internet and the WORLD WIDE WEB. Their Open Marketplace service provides secure payment, real-time CREDIT CARD authorization, sales reports and STOREFRONT management. Their Secure WebServer software supports both S-HTTP and SSL for secure transactions.

open network A capability offered by an EDI VAN by which data can be routed in either direction between a VAN customer and a VAN noncustomer.

open network architecture A policy adopted by the FCC requiring that RBOCs unbundle their services and provide competitors with equal ACCESS to the local exchange.

operational window In EDI, the scheduled times of the day and week when a company is ready to receive EDI transmissions.

optional data element (or segment) In EDI, a DATA ELEMENT (or SEGMENT) which is not required but may be included in a transmission.

order status inquiry An EDI STANDARD TRANSACTION SET used to ask for the status of an order.

org The top-level Internet DOMAIN NAME assigned to non-profit organizations. The domain name of a scholarly or scientific organization ends with a period followed by ORG (eg. CPSR.org).

Orwellian Consistent with the idea that a big brother, most likely the government, is tracking individual usage of the Internet and E-MAIL. Derived from George Orwell's novel, *1984.*

OSI Open Systems Interconnection.

OSI Reference Model The seven-layer architecture for computer internetworking established by the INTERNATIONAL STANDARDS ORGANIZATION (ISO). Starting from layer number one, the layers are: physical, data link, network, transport, session, presentation, and application. Also see TRANSMISSION CONTROL PROTOCOL/INTERNET PROTOCOL.

OTA See OFFICE OF TECHNOLOGY ASSESSMENT.

outsourcing The purchasing of services from outside a company that have been previously performed within a company (e.g. data processing).

▪ P ▪

packet A group of bytes organized for TRANSMISSION over a network and containing control information (most importantly a destination address) and, usually, data. The term is also used loosely to refer to a DATAGRAM or to a block of data at any level of the TCP/IP PROTOCOL STACK. A PACKET generally corresponds to a FRAME as defined in the TRANSMISSION CONTROL PROTOCOL/INTERNET PROTOCOL but may include additional control information added by the network.

packet level filtering The most popular FIRE WALL technology, usually run on a ROUTER. Packet level filtering is able to monitor each TCP/IP PACKET, examine its address and PORT numbers, ensure that traffic is authorized, and route it. It is much easier to implement than an APPLICATION LEVEL GATEWAY because it simply uses the filtering capabilities built into the router. Packet level filtering does not provide full SECURITY for non-TCP/IP packets, is vulnerable to IP SPOOFING, does not provide true user AUTHENTICATION, and provides only limited reporting and auditing.

packet sniffer A software tool that helps maintain, trouble-shoot, and fine-tune LANs and WANs. It observes traffic on network segments, learns network configurations, decodes protocols, delivers statistics, automatically identifies many common network problems, and enables the creation of management reports. See SNIFFING ATTACK.

packet-switched network A NETWORK, such as the Internet, in which data can be exchanged between two computer systems without preliminary exchange of control information or the establishment of a circuit between the two systems. Also known as a connectionless network. See CIRCUIT-SWITCHED NETWORK.

PAN See PRIMARY ACCOUNT NUMBER.

password A sequence of characters paired with a user name, designed to assure that only the authorized person can LOG ON with that particular user name. A good password should be difficult to guess or derive by logic and should be kept secret by the user.

payment cancellation request An EDI STANDARD TRANSACTION SET sent to a BANK to cancel a payment instruction sent previously.

payment card A CREDIT CARD, DEBIT CARD, ATM CARD, SMART CARD, or multiple use BANK CARD combining a number of these cards capabilities. See TRANSACTION CARD.

payment order/remittance advice In the case of DATA SEPARATE FROM DOLLARS an EDI STANDARD TRANSACTION SET that advises a seller that payment has been made. In the case of DATA WITH DOLLARS, an EDI STANDARD TRANSACTION SET sent to a FINANCIAL INSTITUTION which forwards it to the payee and simultaneously effects the transfer of funds.

payment status report An EDI STANDARD TRANSACTION SET reporting on transactions or account balances or both - sent to a customer by a BANK.

payment system A system which enables payments to be made, particularly using digital communications, particularly over closed, private networks such as FED WIRE and AUTOMATED CLEARING HOUSE networks. A payment system authenticates buyers and sellers, validates MESSAGES, provides secure delivery of TRANSACTION information, verifies the buyer's ability to pay, ensures delivery of the merchandise, and has procedures for handling exceptions. For payment systems over OPEN NETWORKS see NETWORK PAYMENT SYSTEM.

pay-per-use Payment for each piece of information or each use of software (including games, videos etc.) as opposed to a one-time fee.

PCA See POLICY CERTIFICATION AUTHORITY.

PCMCIA Personal Computer Memory Card International Association. A standard for the form and interconnection method for CREDIT CARD size enclosed circuit boards that add various peripherals and memory storage, particularly for laptop computers and PERSONAL DIGITAL ASSISTANTS.

PCS See PERSONAL COMMUNICATIONS SERVICES.

PD See PRODUCT DATA.

PDA See PERSONAL DIGITAL ASSISTANT.

PDN See PUBLIC DATA NETWORK.

PEM See PRIVACY-ENHANCED MAIL.

Personal Communications Services (PCS) Wireless cellular services typically provided to low cost telephones and PERSONAL DIGITAL ASSISTANTS and operating at 1.8 to 2.0 Ghz.

personal digital assistant (PDA) A handheld computer, usually with a pen-based user interface and wireless communication capabilities for FAX, data, and paging.

personal identification number (PIN) A secret numeric PASSWORD associated with a BANK account, ATM CARD, or DEBIT CARD, that the account or card holder must enter into a key pad or speak over the phone to demonstrate that the person using the card is authorized to do so. Also known as a SECURITY code or ACCESS number.

PGP See PRETTY GOOD PRIVACY.

Piggyback bar code label A label with two copies of the same bar code. One copy is detached to be read by a SCANNER when the item displaying the label is used or transferred.

PIN See PERSONAL IDENTIFICATION NUMBER.

PIN pad A numeric key pad used to enter a PERSONAL IDENTIFICATION NUMBER when using an AUTOMATED TELLER MACHINE or paying for a retail purchase with a DEBIT CARD.

ping Packet internet groper. A simple IP MESSAGE that requests an echo from another device on the network. While this IP capability is not directly accessible by humans, it is often simulated in the application layer of the TCP/IP PROTOCOL STACK to determine whether a specific network device is reachable over the network.

PIV Personal Identification Verification. Verification of a person's IDENTITY using unique physical characteristics. See also BIOMETRIC AUTHENTICATION.

PKCS See PUBLIC KEY CRYPTOGRAPHY STANDARDS.

PKZIP A DATA COMPRESSION program for DOS and Windows-based computers that uses the.ZIP compression format.

plain text Unencrypted text.

planning schedule An EDI STANDARD TRANSACTION SET commonly used in the automobile industry to provide forecast unit shipments.

PLC Power line carrier. Any technology which permits communication over a wire carrying commercial electric power, inside or outside a home or business.

point of presence (POP) 1) A location where an online service or INTERNET SERVICE PROVIDER offers DIAL-UP PORTS or connects to DEDICATED LINES. 2) A location where a customer or a local telephone company can connect to a long distance carrier.

point-of-sale system (POS) A computer system for retailers that collects product and other information at the time of sale and sends it to a central database for revenue and inventory reporting. Frequently the product is identified by scanning a bar code.

point-to-multipoint A mode of communication in which one party communicates with many, as in telemarketing or home shopping by TV. Broadcast, as in conventional TV and radio, is POINT-TO-MULTIPOINT in which all communication is in a single direction, from the single point to the multiple points.

point-to-point A mode of communication in which two parties communicate only with each other, as in most common voice telephone calls.

Point-to-Point Protocol (PPP) A standard for TCP/IP host-to-network and ROUTER-to-router connections over asynchronous (e.g. dialup) and synchronous lines. A superior alternative to SLIP.

Policy Certification Authority (PCA) Any of the Level 2 organizations in the PEM model which is certified by the INTERNET PCA REGISTRATION AUTHORITY and provides certification to multiple Level 3 Certification Authorities.

POP See POINT OF PRESENCE.

port 1) An interface on a computer or networking device. 2) In TCP, a 16-BIT address uniquely identifying an APPLICATION level service (e.g. Telnet) so that data can be passed between the transport layer of the TCP/IP PROTOCOL STACK and the intended application. Addresses 0-255 are reserved for well-known services and are generally used for communication within a single SERVER. Higher address numbers are assigned as needed for communication between a remote CLIENT and an application on a SERVER. Programs which work over TCP must define which PORT they will use. See SOCKET.

POS See POINT-OF-SALE SYSTEM.

posting An individual MESSAGE sent to a USENET newsgroup.

postmaster The person who has responsibility for ELECTRONIC MAIL in an AUTONOMOUS SYSTEM.

POTS Plain Old Telephone Service. The analog telephone service that is delivered to almost all homes and offices. For comparison, see INTEGRATED SERVICES DIGITAL NETWORK.

PPP See POINT-TO-POINT PROTOCOL.

predefined fixed file A FILE in which fields (or data elements) and records (or segments) conform to predetermined fixed lengths. Such files are generally used to transfer data between an EDI SYSTEM and a business APPLICATION.

preordering transactions EDI STANDARD TRANSACTION SETS which contain information and queries relevant to a sale before an order is placed.

prepaid card A disposable PAYMENT CARD with an electronically stored value which is decremented when funds are transferred from the card to a card reading device.

presence See INTERNET PRESENCE.

Pretty Good Privacy (PGP) A very powerful ENCRYPTION program developed by Philip Zimmerman and freely distributed on the Internet for assuring the PRIVACY of user E-MAIL MESSAGES. PGP uses PUBLIC KEY CRYPTOGRAPHY.

PRI See PRIMARY RATE INTERFACE.

price sales catalog data An EDI STANDARD TRANSACTION SET that updates an ELECTRONIC CATALOG.

primary account number (PAN) On a BANK card, the embossed or encoded number that identifies the ISSUER and the individual account.

primary key An ENCRYPTION key used to protect bulk data. See MASTER KEY.

Primary Rate Interface (PRI) The ISDN interface designed for bulk connection between two circuit switching devices (e.g. between a private branch exchange and a telephone company central office). In the United States the interface includes 23 channels for voice or data and one channel for signaling.

privacy 1) The assurance that a USER is the only person with ACCESS to his or her Internet account and the files in their PC or home directory. 2) In TRANSACTIONS, the assurance that no third party can read the significant information in any of the MESSAGES making up the TRANSACTION.

Privacy-Enhanced Mail (PEM) An Internet standard for providing PRIVACY , AUTHENTICATION, and NON-REPUDIATION over E-MAIL. See POLICY CERTIFICATION AUTHORITY.

private key cryptography Also known as symmetric key ENCRYPTION, an encryption method which requires that both parties to a digital conversation know the same key. The identical key is used for both encryption and DECRYPTION. Conventional CRYPTOG-

RAPHY algorithms using private key include DES, RC4, RC5, and SKIPJACK.

product activity data transaction An EDI STANDARD TRANSACTION SET which relays information on the transfer, sale, or return or a product.

product code An alphanumeric code that uniquely identifies a product.

product data (PD) Information describing every aspect of a product's design, characteristics, and support. See EXCHANGE OF PRODUCT MODEL DATA.

proprietary EDI A non-standard EDI specification developed for the exclusive use of a company and selected trading partners.

proprietary standard A widely-adopted standard created and promoted by a single company. Microsoft Windows is an example.

protocol A description of rules and guidelines that determine how computers and other devices on a network communicate with one another.

protocol stack Also known as a protocol suite, a group of PROTOCOLS that function together to implement a specific communications architecture.

proxy server A SERVER which relays requests for service on behalf of CLIENTS. In some cases the server allows the client to use an application (e.g. ARCHIE) which the client itself does not possess. A PROXY SERVER may also provide FIRE WALL SECURITY between the Internet and a private network by examining the content of packets (e.g. for executable software) before forwarding them.

PSTN Public Switched Telephone Network.

public data network (PDN) A wide area network available to companies and/or individuals for a fee. PDNS include the Internet, some bulletin board services, and ONLINE SERVICES such as America Online.

public domain A program or document not protected by copyright, patents, or trade secrets that can be copied, modified, or resold for any purpose without obligation to the developer.

public key cryptography An ENCRYPTION method which permits secure communication between two parties who have never met and who have not communicated a private key in advance. Each party transmits a public key which the other party uses to encode MESSAGES to be sent back to the first party. The MESSAGES cannot be decoded using the public key but only by a private key which is never transmitted. Public key algorithms include RSA (for ENCRYPTION and digital signatures), El Gamal and DSS (for digital signatures but not ENCRYPTION), and Diffie-Hellman.

Public Key Cryptography Standards (PKCS) A widely used industry sponsored standard for incorporating RSA CRYPTOGRAPHY in standard APPLICATIONS. Compatible with PRIVACY-ENHANCED MAIL and with the x.509 standard for digital certificates.

purchase order An EDI STANDARD TRANSACTION SET sent by a buyer to a seller and containing all information needed to process the order.

purchase order acknowledgment An EDI STANDARD TRANSACTION SET that reports on the expected delivery of the product.

purchase order change An EDI STANDARD TRANSACTION SET identifying the original PURCHASE ORDER and containing changes.

purchase order change acknowledgment An EDI STANDARD TRANSACTION SET reporting on AVAILABILITY and expected delivery in reply to a PURCHASE ORDER CHANGE.

purchase-pay cycle The length of time between the creation of a PURCHASE ORDER and the payment for goods or services received.

▪ Q ▪

QR See QUICK RESPONSE.

qualifier code In EDI, a DATA ELEMENT that indicates which of several possible interpretations should be made of the data in a second, generic data element. Qualifier codes are specified in the X12 data element dictionary.

Quick Response An apparel industry strategy to improve response to customer demand by using various technologies to share POS information between retailers, wholesalers, and manufacturers' suppliers and to jointly forecast demand. See VENDOR-MANAGED INVENTORY and VOLUNTARY INTERINDUSTRY COMMUNICATION STANDARD.

quote An EDI STANDARD TRANSACTION SET that provides price and AVAILABILITY for a product.

▪ R ▪

Radio Frequency Data Collection (RFDC) The instantaneous centralized collection of information by means of wireless handheld bar code scanners.

Radio Frequency Identification (RFID) A technology used on a company's premises for tracking the location of trucks, equipment, or any other item by means of radio buttons attached to the equipment and sensors mounted in various locations.

RBOC Regional Bell Operating Company. One of the seven local telephone companies broken off from AT&T in the MODIFIED FINAL JUDGMENT.

RC2 and RC4 Variable-key-size ENCRYPTION functions designed by Ron Rivest as faster and more flexible alternatives to DES. To qualify for quick export key sizes must be limited to 40 BITS (versus a fixed 56 bits for DES).

receiving advice An EDI STANDARD TRANSACTION SET that advises a vendor of the receipt, quantity received, and condition of products received.

reflector A program that forwards all received MESSAGES to all the members of a MAILING LIST. Also known as a mail reflector or mail exploder.

Regulation E A Federal Reserve Board regulation setting procedures, rules, and liabilities for ELECTRONIC FUNDS TRANSFERS, including consumer protection regulations. Two controversies surround Reg E: 1) In 1997, the rules will be extended to hold state agencies that run ELECTRONIC BENEFIT TRANSFER programs responsible for unlimited losses from EBT DEBIT CARD fraud, potentially increasing a state's costs by millions. 2) The Federal Reserve Board has indicated they believe the regulation covers prepaid SMART CARDS. The SMART CARD FORUM holds that prepaid

SMART CARDS are the digital equivalent of cash and should not fall under Reg E.

Reinsurance and Insurance Network (RINET) A global organization that develops EDI standards for reinsurance and provides EDI service to its members.

reliable protocol A PROTOCOL that assures that data transmitted between computers is received without errors. TCP/IP is reliable because, while IP makes no effort to assure reliability, TCP is a reliable protocol which requires acknowledgment of packets sent and retransmits any not received error-free.

remote access Dialup ACCESS to a network or computer system, usually a private LAN, mini-computer or mainframe.

remote host A HOST (computer) accessed over the network using TELNET, FTP, or some other CLIENT APPLICATION.

remote job entry Entering commands from a remote computer or TERMINAL that cause one or more programs to execute on the computer accessed.

remote server A software program running on a remote HOST that responds to requests from CLIENT software, such as a GOPHER or WORLD WIDE WEB BROWSER.

request for quote An EDI STANDARD TRANSACTION SET requesting price and AVAILABILITY from a vendor and supplying all or a significant part of the information needed by the vendor to furnish the quotation.

Requests for Comments (RFC) Working documents of the INTERNET ENGINEERING TASK FORCE which are often implemented and become de facto standards without any formal process.

resource An item on an Internet HOST which is available to a CLIENT on another HOST. The item might be a program, a FILE, or a tool that supports accessing other resources on the network.

resource discovery tool A program that allows users to find and use RESOURCES on the Internet without requiring the user to know in advance the name of the resource they want to access. Examples include ARCHIE, VERONICA, JUGHEAD, WAIS, GOPHER, and the WORLD WIDE WEB.

retrieval In EDI, the obtaining of incoming data from a MAILBOX on a VAN.

RFC See REQUESTS FOR COMMENTS.

RFDC See RADIO FREQUENCY DATA COLLECTION.

RFID See RADIO FREQUENCY IDENTIFICATION.

RINET See REINSURANCE AND INSURANCE NETWORK.

router A device that forwards packets from one NETWORK to another based on network layer information, using algorithms which optimize speed of PACKET delivery, overall network efficiency or some other parameter.

RSA A public key ENCRYPTION system invented by Rivest, Shamir, and Adelman at MIT and marketed by RSA Data Security Inc. RSA supports MESSAGE encryption , DIGITAL SIGNATURES, and DIGITAL CERTIFICATES. Because encrypting with RSA substantially increases the length of a message, RSA is typically not used to encrypt an entire message but is used to encrypt and exchange a PRIVATE KEY which is then used to encrypt the message.

▪ S ▪

S-HTTP See SECURE HTTP.

scanner A device which converts a printed image into digital information which can be transmitted and used to display or print a copy. A BAR CODE reader goes beyond scanning to interpret the bar code image and translate it into alphanumerics. When coupled with an optical character recognition system, scanned text can often be translated into digital text which can be edited on a computer like any other text document.

SDLC See SYNCHRONOUS DATA LINK CONTROL.

sealed-session networking The ENCRYPTION of data transmissions at a low level in the OSI model (between the transport layer and the DATAGRAM layer) to provide data SECURITY over an enterprise network with minimum sacrifice to network performance.

secret key See PRIVATE KEY.

Secure Hash Standard (SHS) A HASH FUNCTION proposed by NIST for use with the DIGITAL SIGNATURE STANDARD and the only part of the CAPSTONE project which has been adopted as a government standard.

Secure HTTP (S-HTTP) A mechanism developed by Enterprise Integration Technologies (EIT) to enable spontaneous, flexible, and secure commercial transactions on the WORLD WIDE WEB by supporting the ENCAPSULATION of MESSAGES and the negotiation of ENCRYPTION algorithms and other parameters between CLIENTS and servers with varying capabilities. S-HTTP uses the HTTP protocol and is limited to usage with WWW SERVERS and BROWSERS. See TERISA SYSTEMS.

Secure Sockets Layer (SSL) A mechanism developed by Netscape to enable spontaneous, secure commercial transactions on the Internet through ENCRYPTION at the transport layer of the TCP/IP

PROTOCOL STACK. SSL can be used to encrypt communication via the WWW, Gopher, Telnet, and other Internet protocols.

Secure Transactions Technology (STT) A system for enabling secure buying and selling over the Internet in development by Microsoft and Visa International.

security See INFORMATION SECURITY.

security domain A set of computing and network resources typically belonging to one organization and for which a single authority establishes a SECURITY policy.

Security Parameters Index (SPI) A 32-BIT value in an IPV6 DATAGRAM header which identifies which previously negotiated KEYS, ENCRYPTION and/or AUTHENTICATION algorithm, block size, and synchronization/initialization vectors are being used in the datagram. See ENCAPSULATING SECURITY PAYLOAD.

security service A physical control, mechanism, policy or procedure that protects computer communications from THREATS. The types of SECURITY service include AUTHENTICATION, ACCESS CONTROL, CONFIDENTIALITY, DATA INTEGRITY, and NON-REPUDIATION.

segment 1) In the TCP/IP PROTOCOL STACK, a MESSAGE to which a transport header has been added by the transport layer. See DATAGRAM. 2) A logically contiguous portion of a LAN which is not divided by a BRIDGE or a ROUTER. 3) In EDI, a set of data elements that make up a logical unit within a TRANSACTION SET (e.g. a line item on an invoice including line item number, quantity, PRODUCT CODE , product description, and price).

segment tag (EDIFACT) or segment identifier (ANSI X12) T h e first DATA ELEMENT in an EDI SEGMENT containing code which uniquely identifies the SEGMENT.

segment terminator A control character used to identify the end of data in an EDI SEGMENT.

sensitive information As introduced in the Computer Security Act of 1987, sensitive information is "any information, the loss, misuse, or unauthorized access to or modification of which could adversely affect the national interest or the conduct of Federal programs, or the PRIVACY which individuals are entitled to under...the Privacy Act, but which has not been specifically authorized...to be kept secret in the interest of national defense or foreign policy." Privacy is the dominant concern.

Serial Line Internet Protocol (SLIP) A standard for TCP/IP POINT-TO-POINT connections over serial lines.

server A computer or a software program that provides services (e.g. mail, FTP, WWW) to CLIENTS over the network upon request.

service provider See INTERNET SERVICE PROVIDER.

session key A PRIMARY KEY used to encrypt data between computers during a single session (e.g. a TELNET or FTP session).

settlement In banking, the accounting process which records the debit and credit positions of two parties (usually two banks) involved in the transfer of funds. Net SETTLEMENT is the SETTLEMENT on an end-of-day net basis between the reserve accounts of two banks to reflect the result of multiple interbank transactions which have taken place during the day through AUTOMATED CLEARING HOUSES. Final settlement occurs when the net of the transactions (credits less debits) are credited to a banks reserve account by a Federal Reserve Bank.

set-top box A device which converts incoming information into signals that can be accepted by a television set. The signals may be analog (traditional cable TV) or digital and may be received over cable, fiber-optics, twisted pair, by satellite, or even by a variety of cellular communications. The box may also deliver information to PCs and other devices in the home. Set-top box connotes the ability to send information from the viewer back to the originator of the signal to enable video-on-demand, home shopping, interactive TV, games, and other services.

SGML See STANDARD GENERALIZED MARKUP LANGUAGE.

shareware Copyrighted software which is available at no charge for trial by the user. The user is expected to send a specified payment if the program is put to use after expiration of the trial period. In return for the registration fee, the developer of the program may provide printed documentation, software support, and notices of upgrades.

shell account A method of dialup ACCESS to the Internet in which a user interface mimics the interface of a HOST that is on the network (UNIX prompt, menu, or graphical user interface) while the users PC or WORKSTATION is actually connected over a telephone line to the HOST. A SHELL ACCOUNT is a middle ground between DIRECT ACCESS and INDIRECT ACCESS.

Shen A proposal from CERN for WORLD WIDE WEB SECURITY that supports AUTHENTICATION and ENCRYPTION of HTTP connections. Shen has been generally superseded by S-HTTP and SSL.

shipment container marking A 128 BAR CODE label on the exterior of a shipping container that identifies the container and provides other information.

shipment information transaction An EDI STANDARD TRANSACTION SET sent by a carrier in response to a SHIPMENT INQUIRY TRANSACTION.

shipment inquiry transaction An EDI STANDARD TRANSACTION SET sent by a shipper or by the consignee requesting information on a shipment.

shipping transactions A group of EDI STANDARD TRANSACTION SETS related to shipment of goods.

SHS See SECURE HASH STANDARD.

Simple Mail Transport Protocol (SMTP) The worldwide de facto ELECTRONIC MAIL messaging standard. While it facilitates mail between the Internet and other networks, it is limited to ASCII characters. See MIME and PEM.

skimming Copying electronic card data from one PAYMENT CARD to another.

Skipjack The ENCRYPTION algorithm designed by NSA as part of CAPSTONE for encryption between phones, MODEMS, and faxes. It uses an 80-BIT KEY to encrypt and decrypt 64-bit blocks of data. It may be more secure than DES since it uses 80-bit keys and scrambles the data for 32 steps (or rounds) vs. 56-bit keys and 16 rounds for DES. Some details of Skipjack's encryption algorithm are a U.S. government secret.

SKU See STOCK-KEEPING UNIT.

SLAC See SUBSCRIBER LINE ACCESS CONTROLLER

SLIP See SERIAL LINE INTERNET PROTOCOL.

smart card 1) A credit card-sized device implanted with either computer memory chips or computer processors and used for a variety of applications, such as financial debit/credit transactions and computer security. Some smart cards can hold DIGITAL CASH. They may be either reusable cards or PREPAID CARDS and may or may not require the entry of a cryptographic KEY in a card reading device, such as an ELECTRONIC WALLET to transfer the digital cash. Most smart cards will conform to ISO 7816, the international standard for plastic cards containing integrated circuits. See also SUPER SMART CARD, ELECTRONIC PURSE, and MONDEX. 2) In EDI, a floppy disk accompanying a DIRECT STORE DELIVERY which contains the invoice for the delivery. After comparison to the corresponding order on the store's computer, and correction of either company's records if necessary, the invoice is transferred to the store's computer.

Smart Card Forum A nonprofit organization which promotes the development and acceptance of SMART CARD applications.

Smart House A concept backed by the National Association of Homebuilders using a proprietary communications system for monitoring and controlling home energy use that links with a utility-customer communications network based on UCA. See CEBUS.

smart meter An electricity meter (and by extension, a gas and or water meter) that permits communication of meter readings over power line carrier or other means.

SMDS See SWITCHED MULTIMEGABIT DATA SERVICE.

SMTP See SIMPLE MAIL TRANSPORT PROTOCOL.

snail mail Postal Service mail.

sniffing attack An attempt to obtain the content of data transmissions by using a PACKET sniffer or other device at a ROUTER in the network carrying the transmissions.

socket The concatenation of a TCP or UDP PORT number and an IP ADDRESS to uniquely identify a specific APPLICATION service running on a specific SERVER. To a programmer, the SOCKET identifies a communications path. See PORT.

sockets library A set of tools that simplify the development of APPLICATIONS over TCP or UDP.

SOHO Small office/HOME OFFICE The market for computers, office equipment, and services among small offices and home offices, grouped together because they tend to have similar buying patterns.

SONET See SYNCHRONOUS OPTICAL NETWORK.

spamming Mass distribution of unwanted information (usually commercial) to Internet users and newsgroups. To CARPET BOMB is to spam to multiple USENET groups. See MAIL BOMB.

SPI See SECURITY PARAMETERS INDEX.

SprintLink A public data network offered by Sprint which provides Internet ACCESS across the U.S. Usage is not constrained by an ACCEPTABLE USE POLICY.

SSL See SECURE SOCKETS LAYER.

stand alone An adjective applied to a computer system that performs its complete function without connecting to any other computer

Standard General Markup Language (SGML) A language that identifies sections of ASCII text documents (title, footnote, etc.) so that the documents can be transferred between computers and

APPLICATIONS while preserving the general appearance of and logical relationships between the sections.

standard transaction set An EDI TRANSACTION SET that conforms to a company or industry standard, or the ANSI ASC XI2 standard for EDI.

statement An EDI STANDARD TRANSACTION SET that summarizes invoices previously sent over a month or other defined time.

status details reply transaction An EDI STANDARD TRANSACTION SET transmitted from a carrier in reply to a SHIPMENT INQUIRY TRANSACTION set.

STEP See EXCHANGE OF PRODUCT MODEL DATA.

stock item A product stocked by a vendor and having specifications sufficiently well-defined that the product can be unambiguously described by a PRODUCT CODE.

stock-keeping unit (SKU) A number used by a company to specify a unique product received from a vendor (i.e. a unique product specification, not an individual item).

store and forward A widely used method of substantially improving the efficiency of a WIDE AREA NETWORK, a value added network, or the Internet by storing complete MESSAGES at an intermediate point or points and sending them further toward their destination when network capacity is available.

stored value card An ELECTRONIC PURSE. Also see PREPAID CARD.

storefront The representation of a company, with its products and services, on an ONLINE SERVICE or the Internet. Ideally a storefront can close a sale and take an order.

stow plan transaction An EDI STANDARD TRANSACTION SET which provides information on the specific location of a shipment in a ship or barge.

STT See SECURE TRANSACTIONS TECHNOLOGY.

subscriber 1) Unless specifically identified as an E-MAIL or USENET subscriber, a person who has ACCESS to most Internet resources (e.g. TELNET, FTP, WORLD WIDE WEB), and generally a person with a unique user name and PASSWORD. 2) A telephone company customer.

subscriber line access controller (SLAC) A device at a phone company central office that sends and receives signals from a customer electrical meter without interfering with phone traffic.

super smart card A SMART CARD with a display, a key pad, and batteries.

surfing the net User exploration of the Internet in search of interesting or entertaining resources, rather than in pursuit of specific information.

Switched Multimegabit Data Service (SMDS) A PACKET-SWITCHED NETWORK service offered by the telephone companies which can support data, full-motion video, and MULTIMEDIA at speeds up to 150MBPS. See also CONNECTIONLESS BROADBAND DATA SERVICE.

symmetric key encryption See PRIVATE KEY CRYPTOGRAPHY.

Synchronous Data Link Control (SDLC) A BIT-oriented PROTOCOL used for POINT-TO-POINT TRANSMISSION at the data link layer in IBM's System Network Architecture (SNA) and other environments. Also see HDLC.

Synchronous Optical Network (SONET) An ANSI standard for high-BANDWIDTH (up to 2.5 Gbps) TRANSMISSION over optical fiber. The telephone companies expect to use SONET to support SMDS and ATM across the U.S.

synchronous transmission The TRANSMISSION of BITS over a network where the sender and receiver have synchronized clocks. Synchronous transmission is usually more efficient than asynchronous because control information is normally placed around complete MESSAGES rather than around each character.

sysop System operator The person who maintains and operates a bulletin board service or a forum/section on an ONLINE SERVICE.

system   A computer system consisting of hardware and software. By extension, a network of computer systems.

· T ·

T1 A 1.544MBPS telephone line, often used to carry Internet traffic.

T3 A 45MBPS digital wide area network service offered by telephone companies, often used to carry national Internet traffic.

TCP See TRANSMISSION CONTROL PROTOCOL.

TCP/IP See TRANSMISSION CONTROL PROTOCOL/INTERNET PROTOCOL.

TDCC See TRANSPORTATION DATA COORDINATING COMMITTEE.

TDMA See TIME DIVISION MULTIPLE ACCESS.

Technology Reinvestment Program A program administered by ARPA which funds joint research with private companies.

TECnet Technologies for Effective Cooperation Network, a network which links MANUFACTURING TECHNOLOGY CENTERS over the Internet.

Telecenter A project under development by Swiss BANKS to provide HOME BANKING within Switzerland superior to that currently available using VIDEOTEX.

telecommunications Voice, data, or other communications using coded signals (analog or digital) over a TRANSMISSION medium such as wire, fiber optics, radio, infra-red, or other electromagnetic means.

telecommuting The arrangement by which an employee works at home or in a remote office, communicating with his employer's office by telephone and through a data connection from a personal computer or TERMINAL to the employer's network.

teleconference A conference of parties in different locations by means of voice or video communications.

teleputer A theoretical home appliance that combines the PC, the television, and the telephone with sophisticated software and high-speed interactive data communications.

Telnet The standard INTERNET TERMINAL emulation PROTOCOL that permits CLIENTS to log in to remote computers.

Terisa Systems A company jointly owned by EIT, RSA Data Security, NETSCAPE, IBM, America Online, and CompuServe that is packaging S-HTTP and SSL together so that browsers can support both and servers can elect which to use.

terminal A keyboard and a display screen with no local processing or storage that links to a nearby or distant computer system.

text file A FILE which contains only ASCII characters and which can be read on a computer screen by a human. (See also ASCII).

thread In a newsgroup, a succession of related FOLLOW-UP MESSAGES.

threat In computer and NETWORK SECURITY, a person, event, thing or idea which may damage the CONFIDENTIALITY, INTEGRITY, AVAILABILITY, or LEGITIMATE USE of an asset such as a computer system, a network, or information.

Time Division Multiple Access (TDMA) A cellular technology that allows separate transmissions to share a channel (i.e. radio frequency) by allocating the entire BANDWIDTH of the channel to each one of them in turn. (See also CDMA).

token An AUTHENTICATION tool that sends and receives challenges and responses during an authentication process. Tokens are usually hand-held devices similar to calculators but may be CLIENT based software.

Token Ring A LOCAL AREA NETWORK technology developed by IBM.

TQM Total Quality Management.

traceability The ability to determine what individual or company paid certain funds over the network or using a SMART CARD.

Traceability reduces PRIVACY but enhances the ability to identify and prosecute criminal activity.

trading partner Any company with which an organization regularly does business, including the organizations BANKS.

transaction 1) The exchange of money for goods or services, generally used to include the dialogue, negotiation, and follow-up surrounding the exchange. 2) Any event taking place between two parties which changes the net worth or financial position of either party (usually both). 3) The advance of funds, as in a CREDIT CARD transaction. 4) An activity affecting the balance of a deposit account, such as a deposit or withdrawal.

transaction card A credit, debit, or other card which enables transactions, usually by uniquely identifying an account, providing a means of verifying the IDENTITY or the user, or by transferring DIGITAL CASH.

transaction cost The cost of completing a TRANSACTION. Depending on the type of TRANSACTION, the cost comprises the cost of granting credit, the cost of guaranteeing payment, and handling costs. For a $70 transaction, the estimated transaction cost by type of financial instrument is $1.50 by CREDIT CARD, $.25 by DEBIT CARD, $1.00 by check, $.75 by cash (dollar bills), and $2.25 by coin. For transactions handled over the Internet, the transaction cost for each financial instrument (or its digital equivalent) is lower.

transaction level acknowledgment An EDI MESSAGE sent by the receiver of an EDI functional group to report on receipt of the group with the option to reject any or all of the transactions contained in the group.

transaction security Complete TRANSACTION SECURITY includes AUTHENTICATION, PRIVACY, DIGITAL SIGNATURES, and MESSAGE INTEGRITY.

transaction set In EDI, the data sent by one trading partner to another that allows the recipient to complete a single TRANSACTION, essentially a complete business document. For example, a transaction set may be a purchase order with multiple line items. A transaction set is enclosed in a FUNCTIONAL GROUP for transmission. The three parts of a transaction set are header, detail, and summary. Each part of the transaction set is made up of SEGMENTS that conform to a predefined standard.

translation software In EDI, a program that converts data between an EDI standard format and a pre-defined fixed field format or an APPLICATION data format.

transmission The transfer of information from one location to another by electronic, electromagnetic, or optical means.

transmission acknowledgment An EDI STANDARD TRANSACTION SET sent by the receiver of an EDI MESSAGE to the sender prior to validation of the data for business purposes.

Transmission Control Protocol (TCP). The PROTOCOL for the transport layer (layer four) of the TCP/IP protocol suite. TCP is a connection-oriented protocol which provides reliability by handling error detection and correction, flow control, resequencing of SEGMENTS and removal of duplicate segments. It adds a transport header to MESSAGES received from the layer above it, the application layer (layer five), to form segments and pass them to the INTERNET PROTOCOL in the network layer (layer three). See TRANSMISSION CONTROL PROTOCOL/INTERNET PROTOCOL.

Transmission Control Protocol/Internet Protocol (TCP/IP) The common name for a suite of protocols developed for the INTERNET but used in private networks as well. In contrast to the seven layer OSI model, a TCP/IP stack has only five layers: physical, datalink, internet(or network), transport, and APPLICATION.

transparent An entity, usually software, which is present but which is not visible to the USER. Some WWW BROWSERS include TCP/IP communications software which is TRANSPARENT because the user never interacts with it.

Transportation Data Coordinating Committee (TDCC) The group that developed EDI standards specific to the transportation industry.

Triple DES A stronger variation of the DATA ENCRYPTION STANDARD (DES) which encrypts 64-BIT blocks with one KEY, encrypts the result with a second key, and encrypts that result with either the first key again or an entirely different key.

Trojan horse A seemingly benign computer program which contains software designed to disrupt the computer it runs on and possibly destroy data.

TRP See TECHNOLOGY REINVESTMENT PROGRAM.

trust hierarchy An organizational structure which assures the validity of DIGITAL CERTIFICATES by tracing the CERTIFICATE ISSUING AUTHORITY up a hierarchy to a party trusted by the recipient of the DIGITAL CERTIFICATE. See INTERNET PCA REGISTRATION AUTHORITY and PEM.

two-factor authentication AUTHENTICATION of a remote USER based upon something the user knows (e.g. a PASSWORD) and something the user owns (e.g. a number or algorithm embedded in a device such as a TOKEN). Using an ATM CARD and a PIN at an AUTOMATED TELLER MACHINE to gain ACCESS to an account is an example of TWO-FACTOR AUTHENTICATION. See CHALLENGE/RESPONSE.

· U ·

UCA Utility Communications Architecture (UCA) An OSI-based PROTOCOL set that allows information exchange between any communications systems within a utility and facilitates exchanges between utilities. A subset of UCA can be used for communication between a utility and its customers.

UCC See UNIFORM CODE COUNCIL.

UCS See UNIFORM COMMUNICATION STANDARD.

UDP User Datagram Protocol. An alternative to TCP which provides faster communication overall and uses less network BANDWIDTH. In contrast to TCP , UDP is an unreliable service. There is no guarantee that a MESSAGE will be delivered and all data detection and correction must be handled by the APPLICATION layer.

UG Utility Gateway. An interface between various communications media, smart appliances, and intelligent appliances on a customer's premise and a utility using the UCA interface.

UN/ECE The United Nations/Economic Commission of Europe which administers and controls the EDIFACT standard.

Uniform Code Council (UCC) The organization which administers UPC symbols, other retail bar codes and some EDI standards. The UCC assigns UPC vendor numbers.

Uniform Commercial Code A set of laws adopted in all states (with some exceptions in Louisiana), the District of Columbia, and the Virgin Islands to bring uniformity to laws governing the sale of goods, banking transactions, secured transactions in personal property, and other commercial transactions.

Uniform Communication Standard (UCS) The grocery industry EDI standard.

uniform resource locator (URL) A standardized method of identifying any document or RESOURCE on the Internet. For example http://www.svi.org/SVI/events.html indicates using HTTP to get the document events.html from the directory SVI on the SERVER named www.svi.org. The WWW makes extensive use of URLs.

Universal Product Code A numeric code for retail goods that identifies a specific product (e.g. by brand, size, and type). It contains a six-digit code assigned by the UCC to identify the manufacturer, a six-digit code assigned by the manufacturer to uniquely identify the product, and a two-digit check code. See BAR CODE.

universal service A guideline incorporated into the NII and based upon a goal of the Communications Act of 1934 which was, "To make available, so far as possible, to all people of the United States, a rapid, efficient, nation-wide and world-wide wire and radio communication service with adequate facilities at reasonable charges." In both cases the intent has been to assure access to the communication infrastructure for parties (such as rural households and businesses) where the cost of providing access is higher than it is for other parties.

UNIX A multitasking and usually multithreading operating system widely used in engineering and technical APPLICATIONS.

UPC See UNIVERSAL PRODUCT CODE.

upload To transfer data from a CLIENT to a SERVER or to a larger computer over a network, including the Internet.

URL See UNIFORM RESOURCE LOCATOR.

US Customs manifest An EDI STANDARD TRANSACTION SET that provides the customs department with information on a shipment being imported.

usage-based pricing See METERING and PAY-PER-USE.

usage designator In an EDI SEGMENT, a code that indicates whether use of a DATA ELEMENT is mandatory, optional, or conditional. See CONDITIONAL DATA ELEMENT.

USENET An informal group of bulletin boards and discussion groups available on the Internet.

user 1) A person using a computer, an APPLICATION, or a NETWORK. 2) An Internet SUBSCRIBER. See MATRIX.

user-friendly Easily used by a USER, particularly one who is not familiar with the APPLICATION or with computers and software in general.

uuencode A UNIX program that encodes a binary FILE into ASCII so that it can be transmitted over Internet E-MAIL or USENET. The recipient uses the uudecode program to decode the FILE.

· V ·

V.42 An error correction PROTOCOL for use in modems.

V.42bis A common compression standard used in modems and capable of DATA COMPRESSION at up to a 4:1 ratio.

V.Fast A modulation PROTOCOL for MODEMS used prior to adoption of the V.34 protocol. Most V.Fast modems can be upgraded to V.34.

V.xx and V.xx bis See MODEM.

VAB See VALUE ADDED BANK.

Value Added Bank A BANK that is both an intermediary and a network for EDI transactions.

Value Added Network A PACKET-SWITCHED NETWORK that offers special services such as PROTOCOL conversion and data STORE AND FORWARD. See EDI VAN.

VAN See VALUE ADDED NETWORK and EDI VAN.

variable-length field In EDI, a DATA ELEMENT whose length varies with the amount of data contained in the field.

variable-length file In EDI, a FILE in which data elements are variable length fields.

VDT See VIDEO DIAL TONE.

vendor-managed inventory A business arrangement in which a vendor restocks a retailer's inventory without specific orders from the retailer but conforming to an agreed upon model.

Veronica An Internet service that allows keyword searching of GOPHER menu items.

VICS See VOLUNTARY INTERINDUSTRY COMMUNICATION STANDARD.

Video Dial Tone (VDT) A transport service offered by a company that provides no part of the program content. VDT represents an opportunity for telephone companies to earn revenue from cable TV without owning TV program content.

Video on Demand (VOD) A service which allows customers to select specific movies or other shows to be delivered to their home over cable whenever they request.

video conferencing Conferencing between at least two locations by transmitting video and sound of the participants to each other over any medium.

Videotex A now largely obsolete method of sending and displaying text and graphics on PCs, specialized terminals, and TV sets.

virtual Seemingly present but actually not there, as in virtual reality.

virtual corporation A term created by William Davidow for a company that delivers a VIRTUAL PRODUCT.

virtual development Product development by geographically separated team members using a network (e.g. the Internet) to substitute for relocation, travel, and the physical delivery of information.

virtual library A RESOURCE DISCOVERY TOOL that sorts WORLD WIDE WEB resources by subject.

virtual product A product that can be customized for each customer order and delivered almost instantaneously in many varieties.

Virtual Reality Modeling Language A language which supports creation and downloading of three dimensional images from a WEB SERVER to a BROWSER capable of displaying and interacting with the three dimensional image.

virtual storefront See STOREFRONT.

virus A usually malicious computer program which is designed to replicate itself on computer disks and over computer networks, sometimes destroying data and disrupting operations of host computers.

VMI See VENDOR-MANAGED INVENTORY.

VOD See VIDEO ON DEMAND.

Voluntary Interindustry Communication Standard An EDI committee concerned with the retail apparel industry.

VRML See VIRTUAL REALITY MODELING LANGUAGE.

· W ·

W3 See WORLD WIDE WEB.

W3C and W3Org See WORLD WIDE WEB Consortium.

WAIS See WIDE AREA INFORMATION SERVICE.

WAN See WIDE AREA NETWORK.

Warehouse Industry Network Standard (WINS) An EDI standard for transactions between manufacturers and public warehouses. It is designed to be compatible with the UCS standard so that information can be exchanged directly between warehouses and retailers.

warm card A bank card with restricted usage (e.g. an ATM CARD permitting deposits but not withdrawals).

Web The WORLD WIDE WEB.

Web page Text and graphics sent to a Web BROWSER by a WEB SERVER which may or may not fill more than one computer screen but all of which can be viewed by scrolling.

Web server A SERVER supporting one or more WEB SITES which supplies WEB PAGES to Web BROWSERS upon request.

Web site One or more interlinked WEB PAGES controlled by a single organization and linked to a single HOME PAGE.

wet signature As opposed to a DIGITAL SIGNATURE or ELECTRONIC SIGNATURE, the physical signature of a person using ink.

Wide Area Information Service (WAIS) A system for looking up information in indexed databases and libraries across the Internet.

wide area network (WAN) A data communications NETWORK connecting two or more locations of a single organization or multiple organizations with BRIDGES or ROUTERS.

Windows 95 A new version of Microsoft's Windows product which will include easy ACCESS to the forthcoming MICROSOFT NETWORK.

101

WINS See WAREHOUSE INDUSTRY NETWORK STANDARD.

WinSock A specification for development of UNIX-like sockets in TCP/IP PROTOCOL STACKS and APPLICATIONS that run on Microsoft Windows, eliminating the need for APPLICATION developers to implement their own TCP/IP protocol stacks.

wire transfer The payment of funds, usually a large amount, by sending instructions over wire or telephone. Federal Wire and automated clearing houses are examples of wire transfer payment systems.

workflow software Software used on a network to route work between individuals in pre-configured patterns in order to improve efficiency. See GROUPWARE and ELECTRONIC FORM.

workstation A powerful desktop computer, usually running UNIX. As PCs become more powerful the line between PCs and workstations is blurring.

World Wide Web A rapidly increasing set of Internet servers that provide HYPERTEXT services to CLIENTS running World Wide Web BROWSERS such as MOSAIC. The Web features open standards, UNIFORM RESOURCE LOCATORS, and a modular SERVER architecture that supports gateways to back end services such as databases. The Web is on its way to become the worlds largest online information source and transaction vehicle.

World Wide Web Consortium (W3Org) The combination of the MIT W3C at MIT and the Euro W3C based at CERN. The W3Org develops and promotes standards for the www.

worm A notorious self-replicating program introduced on the Internet in November 1988 which disabled over 1,200 computers running certain versions of UNIX. Damages caused by the worm led to the founding of CERT.

WWW See WORLD WIDE WEB.

· X ·

X.25 A ITU-TSS standard for TRANSMISSION of data over analog telephone lines at speeds up to 64Kbps with robust error correction capability.

X.400 '88 OSI E-MAIL Privacy and Authentication Recommendations

X.400 A family of protocols defining the ITU-TSS/ISO messaging service (E-MAIL).

X.500 A family of protocols defining the ITU-TSS/ISO directory service.

X.509 An ISO standard for DIGITAL CERTIFICATES. Version 3 adds extensions to improve administration of the certificates.

X9.F.1 ANSI public key ENCRYPTION standards for the financial industry

X12 The ANSI standards that specify architecture and syntax rules for the variable length multi-industry EDI STANDARD TRANSACTION SETS and the format and content for business transactions to be converted into EDI.

X12 data element dictionary The document which provides the definitions and attributes of the data elements in XI2 STANDARD TRANSACTION SETS, as well as the values that can be used in each DATA ELEMENT and the meaning of each value.

xiwt Cross industry working team. A multi-industry coalition committed to defining the architecture and important technical requirements for a powerful and sustainable national information infrastructure.

· Z ·

'zine A magazine designed for and distributed over the Internet, usually with a weird perspective, content, and format.

Electronic Commerce Web Sites
and Their Uniform Resource Locators

3Com Home Page
http://www.3com.com/

About the www-buyinfo Mailing List
http://www.research.att.com/www-buyinfo/about.html

Adobe Systems Incorporated Home Page
http://www.adobe.com/

Apple Computer, Inc.
http://www.apple.com/

AT&T Home Page
http://www.att.com/

Banking and Finance Home Page
http://www.euro.net/innovation/FinanceHP.html

Banking in the WWW
http://www.wiso.gwdg.de/ifbg/banking.html

Betsi's Home Page
http://info.bellcore.com/BETSI/betsi.html

BizNet Technologies Home Page
http://www.biznet.com.blacksburg.va.us/

BizWeb
http://www.bizweb.com/

Business and Commerce
http://www.einet.net/galaxy/Business-and-Commerce.html

Business:Miscellaneous
http://www.yahoo.com/Business/Miscellaneous

Checkfree Home Page
http://www.checkfree.com/

CMP Interactive Media Group
http://www.wais.com/techweb/img/current/default.html

CommerceNet Home
http://www.commerce.net/

Commercial Services on the Net
http://www.directory.net/

Computer Literacy Bookshops Home Page
http://www.clbooks.com/

Computer-Mediated Marketing Environments Home Page
http://colette.ogsm.vanderbilt.edu/

CoroNet Systems
http://www.coronet.com/

CyberCash Home Page
http://www.cybercash.com/

DigiCash home page
http://www.DigiCash.com/

Digital's Commercial Services Home Page.
http://www.service.digital.com/

DISA HOME PAGE
http://www.disa.org/

Ecash
http://www.digicash.com/ecashinfo/detail-info.html

ECIF Home Page
http://waltz.ncsl.nist.gov/ECIF/ecif.html

EDI Help Desk
http://www.well.com/www/unidex/

Electronic Cash, Tokens and Payments in the National Information Infrastructure
http://ganges.cs.tcd.ie/mepeirce/Project/Pro/ToC.html

Electronic Commerce
http://www.zurich.ibm.ch/Technology/Security/sirene/outsideworld/ecommerce.html

Electronic Commerce Association
http://www.globalx.net/eca/

Electronic Commerce Network (ECNet)
http://enws324.eas.asu.edu/cimsys/projects/ecnet/ecnet.html

Electronic Commerce Resources
http://www.premenos.com/Resources/guide.html

Electronic Data Interchange Standards
http://www.premenos.com/standards/EDIStandards.html

Electronic Storefronts
http://www.cs.colorado.edu/homes/mcbryan/public_html/bb/15/summary.html

Enterprise Integration Technologies
http://eit.com/

Fast EDGAR Mutual Funds Reporting
http://edgar.stern.nyu.edu/mutual.html
http://www.oak-ridge.com/topibrp1.html

Financial Encyclopaedia
http://www.euro.net/innovation/Finance_Base/Fin_encyc.html

FINWeb Home Page
http://riskweb.bus.utexas.edu/finweb.html

First Virtual (TM) Home Page
http://www.fv.com/

FSTC Home Page
http://www.llnl.gov/fstc/

Global Network Navigator Home Page
http://gnn.com/gnn/gnn.html

HTTP Security group of W3C
http://info.cern.ch/hypertext/WWW/Security/Overview.html

IBC: Internet Business Center
http://tig.com/IBC/

IBC: Marketing on the Internet
http://tig.com/IBC/White/Paper.html

Index of /cpsr/privacy/epic/
http://cpsr.org/cpsr/privacy/epic/

Information Innovation's Home Page
http://www.euro.net/innovation/WelcomeHP.html

Information on CARI
http://www.netresource.com/itp/cari.html

information superhighway from FOLDOC
http://clinton.ai.mit.edu/white-house-publications/1993/12/1993-12-20-background-on-information-superhighway.text

InfoSeek Home Page
http://www.infoseek.com/Home

Interesting Business Sites on the Web
http://www.rpi.edu/~okeefe/business.html

International Business Machines
http://www.ibm.com/

Internet & Media Research Center
http://www-bprc.mps.ohio-state.edu/cgi-bin/hpp?mrc.html

Internet Business Connection
http://www.charm.net/~ibc/

Internet Infoguide
http://www.internic.net/infoguide/wusage/

Inet-Marketing mailing list archive by thread
http://maigret.popco.com/hyper/inet-marketing/

InterNIC Directory and Database Services - Page 1
http://www.internic.net/ds/dspg01.html

MarketNet - The Electronic Marketplace - Main Entrance
http://mkn.co.uk/

marketplaceMCI
http://www2.pcy.mci.net/marketplace/index.html

MecklerWeb Home Page
http://www.mecklerweb.com/demo.html

Mondex Home Page
http://www.mondex.com/mondex/home.htm

NCSA httpd: FAQ
http://hoohoo.ncsa.uiuc.edu/docs/FAQ.html

NECX On-Ramp
http://necxdirect.necx.com/

NetBill Project Home Page
http://www.ini.cmu.edu/netbill/

NetMarket Homepage
http://netmarket.com/nm/pages/home

Network Payment Mechanisms and Digital Cash
http://ganges.cs.tcd.ie/mepeirce/project.html

NIST WWW - Home Page
http://www.nist.gov/

NSNS /MouseTracks/ - Marketing Activities and Resources
http://nsns.com/MouseTracks/

NSNS /MouseTracks/ Hall of Malls
http://nsns.com/MouseTracks/HallofMalls.html

NSNS /MouseTracks/ The List of Marketing Lists
http://nsns.com/MouseTracks/tloml.html

Oak Ridge Research Homepage
http://www.oak-ridge.com/orr.html

Oak Ridge Research's Selected Internet Business Resources
http://oak-ridge.com/topibrpl.html

O'Reilly & Associates, Inc.
http://www.ora.com/

On Security
http://home.mcom.com/info/security-doc.html

OneServer HOME PAGE
http://www.connectinc.com/

Open Market Home Page
http://www.openmarket.com/

Payment mechanisms designed for the Internet
http://ganges.cs.tcd.ie/mepeirce/Project/oninternet.html

Premenos' Electronic Commerce Resource Guide
http://www.premenos.com/

Reinventing Business In An Electronic Age
http://www.openmarket.com/reinven2.html

Relevant Internet Resources
http://www.rpi.edu/~okeefe/infosys/IJ/internet.html

RSA Data Security, Inc.'s Home Page
http://www.rsa.com/

Rutgers WWW-Security Index page
http://www-ns.rutgers.edu/www-security/index.html

SBA: Small Business Administration Home Page
http://www.sbaonline.sba.gov/

Sirene Home Page
http://www.zurich.ibm.ch/Technology/Security/sirene/index.html

SPRY Home Page
http://www.spry.com/

Spyglass Home Page
http://www.spyglass.com/

Tandem Computers - WWW Home Page
http://www.tandem.com/

Telecom Information Resources
http://www.ipps.lsa.umich.edu/telecom-info.html;mark=286,59,62

Terisa Systems
http://www.terisa.com/

The CAFE project
http://www.digicash.com/cafe/cafe.html

The OpenNet Database
http://www.doe.gov/html/osti/opennet/openscr.html

The Pipeline New York
http://www.pipeline.com/

The Web Developer's Journal
http://www.awa.com/nct/software/eleclead.html

The Whole Internet Catalog
http://gnn.com/cgi-bin/imagemap/HOME?67,95

The World Wide Web Initiative: The Project
http://info.cern.ch/

Thomas Ho's favorite Electronic Commerce WWW resources
© 1994, 1995 Thomas I. M. Ho
http://www.engr.iupui.edu/~ho/interests/commmenu.html

USWeb
http://www.emi.com/usweb/

VISA Home Page
http://www.visa.com/visa/OneCard.html

W3 Search Engines
http://cuiwww.unige.ch/meta-index.html

Web Payment System Technical Information
http://www.openmarket.com/about/technical/payment/

WebCrawler Searching
http://webcrawler.cs.washington.edu/WebCrawler/WebQuery.html

WebMall - Digital's Electronic Mall
http://www.service.digital.com/html/emall.html

Welcome to GE Information Services Inc.
http://www.ge.com/geis/index.html

Welcome To NetChex
http://www.netchex.com/

Welcome to Netscape
http://www.netscape.com/

Welcome To Oracle
http://www.oracle.com/

Welcome to Sun Microsystems
http://www.sun.com/

Welcome To Surety!
http://www.surety.com/

Welcome to Verity
http://www.verity.com/

What's New With NCSA Mosaic
http://www.ncsa.uiuc.edu/SDG/Software/Mosaic/Docs/whats-new.html

World-Wide Web Home
http://info.cern.ch/

WWW Business Yellow Pages
http://www.cba.uh.edu/ylowpges/ylowpges.html

www-buyinfo Home Page
http://www.research.att.com/www-buyinfo/

www.disa.org
http://www.disa.org/disa/whatdisa.html

www.tig.com
http://www.tig.com/IBC/

www.tis.com
http://www.tis.com/

WWWW - WORLD WIDE WEB WORM
http://www.cs.colorado.edu/home/mcbryan/WWWW.html

Xerox, The Document Company
http://www.xerox.com/

XIWT Home Page
http://www.cnri.reston.va.us:3000/XIWT/public.html